Theodor Heuss

ANTON DOHRN

A LIFE FOR SCIENCE

Introduction by
Karl Josef Partsch

With a contribution by
Margret Boveri

Translated from the German by
Liselotte Dieckmann

Edited by
Christiane Groeben

Springer-Verlag
Berlin Heidelberg New York
London Paris Tokyo
Hongkong Barcelona
Budapest

Previous German Editions:
1940, 1948 by Rainer Wunderlich Verlag,
Hermann Leins GmbH & Co., Tübingen

With 9 Figures

All Figures from Archives Stazione Zoologica »Anton Dohrn«

Frontispiece: Anton Dohrn. Study in oil by Hans von Marées (1873)

Library of Congress Cataloging-in-Publication Data. Heuss, Theodor, 1884–1963. [Anton Dohrn in Neapel. English] Anton Dohrn : a life for science / Theodor Heuss ; introduction by Karl Josef Partsch ; with a contribution by Margret Boveri ; translated from the German by Liselotte Dieckmann ; edited by Christiane Groeben. p. cm. Translation of: Anton Dohrn in Neapel. Includes bibliographical references and index.

ISBN-13: 978-3-642-76349-6 e-ISBN-13: 978-3-642-76347-2
DOI: 10.1007/ 978-3-642-76347-2

1. Dohrn, Anton, 1840–1909. 2. Stazione zoologica di Napoli – History. 3. Zoologists – Germany – Biography. I. Groeben, Christiane. II. Title.
QL31.D6H4813 1991 591'.092 – dc20 [B] 91-16705 CIP

Softcover reprint of the hardcover 1st edition 1991

Typesetting: Triltsch, Würzburg

31/3145-543210 – Printed on acid-free paper

TABLE OF CONTENTS

IV THE STRUGGLE FOR THE PROJECT

V SECURITY AND ACCOMPLISHMENT

EDITORIAL REMARKS

The present translation of Theodor Heuss' biography of Anton Dohrn is based on the second and last enlarged edition (Tübingen, Rainer Wunderlich Verlag Hermann Leins, 1962). No parts have been omitted. However, in order to answer requirements of both today's readers and historians of science, the following additions have been made. K. J. Partsch has been asked to introduce the author Theodor Heuss and to update the history of the Naples Station. The index of names has been completely revised. The author's notes have been completed where possible and are given according to modern bibliographical standards; they are to be distinguished from the translator's footnotes which are marked with an asterisk. Misspellings of names and dates in the text have been corrected.

The plan for a translation of Anton Dohrn's biography into English dates back to preliminary enquiries in 1980 at Woods Hole about the possible interest in such a translation. Since then, various persons have encouraged me to go on with this project and have contributed in one way or another. These are: Garland E. Allen, Jean Ann Gilder, Heinz Götze, Victor Hamburger, Hans Körner, and Jane M. Oppenheimer. Special thanks go to them all.

Stazione Zoologica »Anton Dohrn« Christiane Groeben
Villa Comunale
80121 Naples, Italy

September 1990

ACKNOWLEDGMENTS

My heartfelt thanks go to my main helpers (by no means the only ones!): Christiane Groeben, Stazione Zoologica »Anton Dohrn«, Naples, who made suggestions for English changes and read the entire book for typing and other errors; to Viktor Hamburger, who translated pages of biological information of which I was more ignorant than the original author Heuss; and to Elizabeth Patton, who patiently typed, retyped, and corrected the whole work.

6906 Columbia Place
St. Louis, MO 63130, USA

LISELOTTE DIECKMANN

INTRODUCTION

ANTON DOHRN
HIS BIOGRAPHER – HIS INTERNATIONAL RESEARCH CENTER

Today's reader of Anton Dohrn's biography, which was written in 1940, is probably left with a taste for more information on the author and on the fate of Anton Dohrn's creation, the Naples Zoological Station. The following is an attempt to satisfy this curiosity.

When Theodor Heuss (1884–1963) was first approached to write a biography of Anton Dohrn, he doubted his ability to do justice to Dohrn's achievements, being neither a biologist nor a professional historian. Such qualms were unfounded, as becomes apparent from reading the book as well as from a closer look at Heuss' own biography and preparation.

An English encyclopedia entry defines Heuss, the first President of the Federal Republic of Germany (1949–1959), as »a liberal publicist and man of letters«. This is true, though hardly exhaustive. Heuss certainly was »liberal«, but with a strong commitment to social issues. As a publicist he dealt widely with general culture, mainly the arts, history, and economics, but also with politics. That he came from a small rural town in Württemberg played an important part in forming his character. His doctoral thesis on *The Viniculture and the Status of Winedressers in the Neckar Valley* focused on the social and economic problems of his native region.

In 1905, while working on his thesis, a close friend suggested he visit the exhibit of Hans von Marées' paintings [1] at Schleissheim Castle. This friend was Wolf Dohrn (1878–1914), second son of Anton Dohrn. More than 30 years later, Wolf's brother Boguslaw was to ask Heuss to write a biography of Anton Dohrn to celebrate the 100th anniversary of his birth (December 29, 1840).

Heuss' career as a publicist started early [2]. At the age of 21 he was already a member of the editorial staff of *Die Hilfe* [The Help], a periodical backing the ideas of Friedrich Naumann (1860–1919), social reformer and tutor of the young Heuss. From 1912–1918 Heuss edited in Heilbronn a local newspaper and a political monthly. Later he went to Berlin as secretary of the *Deutscher Werkbund,* an association promoting art nouveau design. During this period he also started to teach civics at the newly founded School of Politics (Hochschule für Politik). Clearly Heuss did not tend toward specialization, as many young people did at that time; instead he could be defined a generalist.

Many of his activities were political. Before reaching voting age, Heuss had already organized a successful election campaign for Naumann, who entered Parliament as a left wing liberal representative. Two decades later, Heuss sat in Parliament himself. Although appreciating his services, the leaders of his party, many of whom were elderly, for many years assigned him to the role of »crown prince«. Only from the period 1924 to 1928, and from 1930 onwards was he a member of the *Reichstag*, serving on the Berlin City Council in the intermediate period.

In 1932 Heuss published a critical analysis of the political aims of Hitler [3], and in Parliament he vigorously attacked the national socialist movement, but had to concede to the majority of his parliamentary group, that lacked the courage to vote against the act enabling Hitler to establish dictatorial rule.

»Internal emigration« is the term frequently used in Germany for those who remain at a critical distance from the regime in power without openly expressing their antagonism. Such emigration is therefore spiritual rather than physical and is frequently connected with nostalgia for times gone by.

Theodor Heuss can be said to have »internally emigrated«. After having been ousted from Parliament and from his teaching position at the School of Politics, Heuss turned his attention to research on people who had lived for values that by then seemed to have vanished. He was to write the biographies of Friedrich Naumann (1937), the architect Hans Poelzig (1939), the chemist Justus von Liebig (1942), and the industrialist Robert Bosch (1946). It is logical to find the biography of Anton Dohrn in this series.

After the war, Heuss was elected to the Parliament of Baden-Württemberg, where he served as Minister of Culture (1945–46). He was one of the most active members of the Parliamentary Council, entrusted to draft the Fundamental Law of the Federal Republic of Germany (1948/49). He made important contributions to the Preamble and the Catalogue of Human Rights. The name of the state was proposed by him, as was the constitutional structure that elected the President, i.e., the Federal Assembly (Bundesversammlung). In recognition of his achievements, on September 12, 1949, Heuss was elected the first President of the Federal Republic. Although the office held limited powers, Heuss vested it with authority and dignity – not an easy task with the authoritarian Konrad Adenauer as Chancellor. Heuss' patience, persistence, humor, and persuasive charm clearly emerge from the recently published correspondence between the two statesmen [4].

The biography of Anton Dohrn was written more than 40 years ago, primarily for a German public more or less familiar with the scientific and political organization of their country. A postscript to the work contains some brief remarks on the development of the Zoological Station after Dohrn's death in 1909. More information on both aspects may therefore be welcome [5].

The Zoological Station, once defined as »a permanent international congress of zoologists« [6], was financed in part from renting »work tables«. The main support came from Italy, Germany, the United Kingdom, and the United States of America, and continued under Anton Dohrn's successors first Reinhard and then Peter Dohrn. It is therefore appropriate to describe these relations on a country-by-country basis.

1 Italy

Heuss is very cautious in describing the fate of Anton and Reinhard Dohrn's property during and after World War I. He does not mention details such as the seizure as alien property in 1915, followed by confiscation, or the restitution through the then Secretary of Education, the liberal philosopher Benedetto Croce (1866–1952). Nor does he mention the occupation of the Station by extremist Neapolitan students (1922) or the successful action by Reinhard Dohrn in the Italian courts, whose verdict was disregarded by Mussolini. Giovanni Gentile (1875–1944), philosopher, Secretary of Education, and pupil of Croce, agreed to a solution which confirmed the confiscation of the Station, but entrusted the former owner with the task of its administration. The Station was, by this time, an Italian institution. With great diplomatic skill Reinhard Dohrn succeeded in re-establishing and maintaining the spirit of international collaboration in the *Stazione Zoologica* [7]. During the periods of political crisis which followed, Dohrn was supported by his friends among the members of the Administrative Council, composed of Italian scientists and officials.

On the 100th anniversary of Anton Dohrn's birth (1940), a permanent seat in the Station's Administrative Council was assigned to a representative of the founder's family, a street was named in Anton Dohrn's honor, and honorary citizenship was conferred to Reinhard Dohrn.

During the period between the two wars Reinhard Dohrn's position was never questioned, thanks in part to Benedetto Croce, who cushioned any hostility. Reinhard Dohrn was therefore able to continue the work of his father and to guide the Station through all – mostly financial – difficulties to a greater efficiency and to an enhanced international reputation. During these years more than half of the finances for the Station had come from the Italian Government.

When Reinhard Dohrn retired (1954), it was decided that his son, Dr. Peter Dohrn, should succeed him to continue the tradition. The three generations were once characterized by Reinhard Dohrn as »founder, conservator, and innovator«. Indeed, Peter Dohrn undertook far-reaching innovations: he increased the number of tables for guest scientists; he obtained support for important research projects from both the United States and from NATO; he rebuilt the Library with substantial contributions from non-Italian donors. He also remodeled the physiology and biochemistry laboratories and, as a consequence, increased the permanent staff. This led to conflict with the Administrative Council and with the authorities in Rome. In addition, the staff of the Station, supported by the newly founded labor union, demanded the right to perform independent research instead of only providing services and facilities to visiting scientists. This change of policy found favor in Rome.

In 1967, the Board of Administration resigned and as a consequence Peter Dohrn was dismissed as director. Both functions were entrusted to a government commissioner who had a mandate to draft a new statute.

From then on, the de facto international character of the Station diminished, although the two subsequent statutes, approved in 1973 and 1982 respectively, stressed the continuation of this tradition. In the 1982 statute the name of the institute was changed into *Stazione Zoologica »Anton Dohrn«*, while the family lost its seat in the Administrative Council. Today, the Stazione is an Italian public institution under the direct control of the Ministry of Scientific Research [Ministero per l'Università e la Ricerca Scientifica e Tecnologica]. The promotion of international scientific collaboration has not been continued to the same extent.

2 Germany

It comes as no surprise that Theodor Heuss lays so much stress on the relations between Anton Dohrn and his native country. While the Station was the private property of a German national, no other country subsidized it which larger sums than Germany. Between

1879 and 1908, the contributions from German public sources amounted to an average of 33.6% of the whole budget, while Italy contributed 13.9%, and all other nations collectively 16.4%.

When writing the book, sources from Germany may have been more easily available to Heuss than those from other countries, and given his experience as a publisher, he may have assumed that his public would be mainly interested in the relations between Dohrn and his own country. Today's reader may be slightly irritated by the expressions of national pride in the achievements of a co-national, but the tone reflects the period and conditions under which this book was written.

Anton Dohrn preferred to negotiate with the authorities of the Imperial Government; he could not, however, neglect the fact that matters related to science, research, and instruction were primarily a competence of the Länder, the single German states, who stipulated independent contracts for the rent of work tables at Naples.

During the Weimar Republic relations between Italy and Germany were, to begin with, unfriendly and all German contributions had to be disguised as funds from scientific institutions, respecting a balance between Italian and German funding.

Like his father, Reinhard Dohrn objected persistently to attempts to bring the Station under the control of governmental authorities, as occurred for example during the close alliance between Germany and Italy after the Spanish Civil war.

From correspondence with friends, it is clear that Reinhard did not support the Nazi rule; but by openly opposing the regime he would have risked losing badly needed financial support. After Italy had changed allies, Dohrn's personal position was officially acknowledged.

Because of the economic situation after World War II it took some time for scientific relations with the Federal Republic of Germany and, for a short period also with the German Democratic Republic, to be re-established. Theodor Heuss, acting as a kind of »clandestine« Federal Secretary of Culture, lent a hand here. He personally contacted the Secretaries of Culture of the single German states to renew the contracts for the research tables at Naples. In 1956 the Federation contributed to the cost for a new library and also supported the restoration of the Marées frescoes. Until very recently the Deutsche Forschungsgemeinschaft [German Research Council] generously supported the Station's unique library and archival collections. During the period from 1954 to 1966 almost 20% of the Station's budget came from German sources. Since 1966, however, there has been a general gradual reduction of contributions from non-Italian scientific organizations and governments.

3 United Kingdom

The friendly style in which Charles Darwin and Anton Dohrn corresponded between 1860–1880 [8] is characteristic of the relationship with other British scientists, who continuously sent their students to Naples; British contract partners for research tables – unlike those of continental countries – were also individual universities or scientific institutions (Cambridge University, Oxford University, The British Association, The Royal Society). During the difficult period after World War 1, when British scientists hesitated to enter into contact with a former enemy, it was Julian Huxley, the grandson of Anton Dohrn's friend, supporter, and correspondent, Thomas Henry Huxley, who paved the way for a renewal of the close international collaboration between scientists from all nations. He did, however, suggest »some form of international supervision«. The idea of international control either by the League of Nations or by ad hoc organized groups of friends of the Station from different countries, did not survive the takeover of Mussolini's followers. This did not, however, prevent the partners of the Station from renewing or reviving their table contracts, thanks mainly to their friendship with Reinhard Dohrn. The three British tables (Oxford, Cambridge, British Association) were so intensively used, that it was sometimes difficult to find a place for all the table holders.

After World War II, reactions in Great Britain were quite different from the period of hesitation after the first World War. When in October 1943 news reached London that the Station had survived the war, George Parker Bidder (1863–1953), one of the most ardent antagonists of Julian Huxley during the 1920's, wrote to *The Times:*

»The news that the fish are still swimming in the tanks of the Aquarium, with the news I have to-day that the Director (friend of all of us) was still there in charge in August, give grounds for hope that we may yet see in the Station again men of many tongues and more nations working side by side in the fraternity of a common endeavour to understand the living world that surrounds us.« [9]

In March 1944, the Royal Society gave £ 1000 for the maintenance of the *Stazione Zoologica* and subsequently rented, together with the University College, London, the fourth and fifth work tables. During the following years, all tables were fully used, sometimes reaching peaks of 110–120% (man/table/day). In 1956, a campaign to raise funds for the new library also yielded gratifying results. A few years later (1963) the five British tables came under the central administration of the Royal Society and table fees were raised to equal those from other countries. In 1972 the Royal Society provided 3.38% of the budget of the Station – a peak never to be reached again. During the 1970's the number of tables was first reduced drastically and finally given up because of financial cuts on one side and diminished interest in having guest researchers on the other.

4 United States of America

Relations between the United States and the Naples Station began much later than those with the United Kingdom. Although a few American scientists showed early interest in using the Naples facilities, longer-lasting contractual relations were taken up by four institutions only after 1890 (Smithsonian Institution, 1893; Columbia College/University, 1896; Women's Association, 1898; Carnegie Institution, 1902). Two of them, the Columbia and the Women's Table, were renewed after 1924. A third table was rented by the Rockefeller Institute for Medical Research (New York) and named in memory of the famous physiologist Jacques Loeb (1859–1924), who had worked at Naples in 1889 and 1890. The ties between US institutions and the *Stazione Zoologica* were mostly due to the personal efforts of the cytologist and close friend of Anton Dohrn, Edmund Beecher Wilson (1856–1939) from Columbia University. Wilson came to Naples for the first time in 1883. During the early 1920's he actively supported the re-instalment of Reinhard Dohrn as director.

In 1924, the director of the Station had to deal with an extremely difficult question. The *International Education Board* (I. E. B.), founded by John Davison Rockefeller Jr., generously offered substantial subsidies to the Naples institute by renting as many as 20 tables primarily for the use of scientists from countries with non-convertible currencies. Fearing that such an agreement might endanger the Station's international character, typically Dohrn came up with a compromise which allowed him to accept the subsidies without losing the collaboration of the former contracting states [10].

This example clearly indicates that the system of renting research tables not only constitutes a technical means for financing a research institute. According to the concept developed by Anton Dohrn, it also creates a kind of partnership and close long-term co-operation.

In the critical year 1933, the Rockefeller Foundation assigned a substantial sum to Reinhard Dohrn personally which allowed him to help financially the emigration of political refugees from Germany.

After 1945, American subsidies to the Station had more or less the character of an emergency service generously provided for by the US National Research Council and the Rockefeller Foundation in recognition of former merits. The latter also participated in financing the new library.

When the number of American scientists coming to Naples to use the facilities there, increased, the National Science Foundation (founded in 1950) rented several tables (two at first, going up to ten in 1961) and also contributed considerable amounts to the general budget of the Station. The US Public Health Service and various US military organizations conducted important research projects in Naples.

The close collaboration with American institutions came to an end in 1971, when financial support for scientific research in the USA was drastically cut. This coincided with a cooling of the relationship between the Italian authorities and the American scientific communities.

Anton Dohrn's lifetime was a period of continual progress in science. He had created his *Stazione Zoologica* under specific scientific, social, and political conditions and had met with remarkable success as long as these conditions lasted. Fundamental changes occurred during the two World Wars and it was not only the unforeseeable events of international politics that affected his creation. Individual research was increasingly replaced by team work and interdisciplinary projects, while collaborators and assistants no longer accepted paternalistic guidance. In addition, similar institutions have been founded in many countries. Thanks to an understanding of these mutations and in some cases, foresight, the heirs of Anton Dohrn have been able with great flexibility to meet the challenge and continue its service to science for more than half a century.

Frankenstrasse 10
6507 Ingelheim, FRG

KARL JOSEF PARTSCH

1. Theodor Heuss, *Die Begegnung mit Hans von Marées.* In: Bernhard Degenhart, *Hans von Marées – Die Fresken in Neapel.* München, Prestel, 1958, p. 9
2. Hildegard Hamm-Brücher und Hermann Rudolf, *Theodor Heuss. Eine Bildbiographie.* Stuttgart, Deutsche Verlagsanstalt, 1983, pp. 45 ff
3. Theodor Heuss, *Hitlers Weg. Eine historisch-politische Studie über den Nationalsozialismus.* Stuttgart, 1932; 2nd ed. Tübingen, R. Wunderlich, 1968
4. Heuss – Adenauer. *Unserm Vaterlande zugute.* Der Briefwechsel 1948 – 1963. Berlin, Siedler Verlag, 1989
5. For more details see: Karl Josef Partsch, *Die Zoologische Station in Neapel. Modell internationaler Wissenschaftszusammenarbeit.* Göttingen, Vandenhoeck & Ruprecht, 1980
6. Theodor Boveri, *Anton Dohrn. Gedächtnisrede gehalten auf dem Internationalen Zoologen-Kongress in Graz am 18. August 1910.* Leipzig, S. Hirzel, 1910, p. 40
7. *Reinhard Dohrn 1880–1962. Reden, Briefe und Veröffentlichungen zum 100. Geburtstag.* Berlin Heidelberg New York, Springer, 1983
8. *Charles Darwin – Anton Dohrn. Correspondence,* edited by Christiane Groeben. Naples, Macchiaroli, 1982
9. Letter to the editor. *The Times* (Oct. 14, 1943)
10. K. J. Partsch (note [5]), pp. 208 – 210

PREFACE · 1940

When he was nearly 60 years of age, Anton Dohrn started to write the history of his work, the Zoological Station in Naples. This memorial was intended to refer only to the Station rather than to become an autobiography, although the author also had to tell about himself and his plans, adventures, and decisions. The manuscript starts only with the year 1870. But when Dohrn died in 1909 only a torso was available, extending only to the year 1874.

It was the purpose of his widow (who had participated in a practical and very productive way in the early stages of the project in years full of sacrifices) to continue the work and, in a certain way, to reshape it by preceding it with a short introduction about her husband's early years and education. With great conscientiousness, she started to collect or make selections from typical letters; a connecting text served to interpret or characterize them. This was carried out until 1878. For the following years she planned a certain pattern, according to which letters and other material would be used for one or another field of work. Marie Dohrn worked on this at several places – in Naples, Forte dei Marmi, Hellerau (near Dresden), and above all on the family estate, Wydranka, in White Russia. She was living there when the World War started and died there in 1918. Inquiries started after the war resulted in the resigned statement that in the storms of the Russian revolution the Baranowski family's archive had been plundered and burnt, and that not only her notes but also the most important part of Dohrn's correspondence which she used were lost.

The reader of this book will find that in Anton Dohrn's life surprising and improbable events occurred. That was also true for the archive. While it was supposed to have been destroyed in Russia, it led a completely undisturbed, but also unnoticed life in suitcases and boxes in the basement of a former private school at

Hellerau near Dresden. This was a foundation by Anton Dohrn's second son, Wolf, but its potential development was interrupted by the war. There Marie Dohrn had preserved the material. Her visit to her Russian home had been planned to be short, but her 93-year-old father became ill and died a few days before the war broke out, and it was impossible for her to return home. When, in the early summer of 1939, the Hellerau property was sold to a state institution, the lost papers, believed to have been destroyed in Russia, were found, after a quarter of a century.

This was the moment when my friend Boguslav Dohrn asked me to examine the existing material. He was moved by the desire to fulfill the old plan of a complete evaluation, if possible, for the 100th birthday of his father on December 29, 1940, and he suggested that I take over this work. The answer required thorough consideration. To be sure, for decades I had been well acquainted with Dohrn's »atmosphere« (if it can be called thus), by a friendship reaching back in part to my university years, with Anton Dohrn's sons and their numerous tales of their father and grandfather; but I shied away from any claim that I, whose literary work was in completely different fields, could expect an audience in the field of natural science. For in this I am an utter layman, not knowing about biology more than its general problems, a sketch of its history, and never having had a closer relationship to its method. It would appear to me to be an impermissible arrogance if I wanted to show in this field an expert evaluation of individual accomplishments.

For two reasons these doubts were overcome. My study of the material, with the intimate human documents as well as the various official documents confirmed how much Dohrn in all his activities had lived outside the framework of limited specialized science. Quite aside from the fact that, in the course of the decades, he published a fairly voluminous series of phylogenetic studies and with them broached a still incomplete specialized scientific discussion, his person, temperament, willpower, and general education were so rich in typically contemporary and yet individual features that it was a temptation to investigate his nature and achievement. This achievement confronted the onlooker with its unique quality: it can be taken as the first »research institution«. The dialogue between personal enthusiasm and the scientific and political decisions of different nations makes both the study and the relating of it fascinating. Added to this was the encouragement I found among some leading German biologists whom I had asked for advice. Dohrn »should not« be discussed by a specialist. The latter would have the task of examining the questions, working manner, and results of Dohrn's investigations. Professor Alfred Kühn of the Kaiser-Wilhelm-Institut für Biologie, Berlin-Dahlem, is preparing an exhaustive study for the *Mittheilungen aus der Zoologischen Station zu Neapel* on *Anton Dohrn und die Zoologie seiner Zeit* [Anton Dohrn and the zoology of his time]. However, the general opinion was that Dohrn's unusual significance for science could not be discussed exhaustively by a critical analysis of his *Studien*. The encouragement of men who had worked personally under Dohrn, and who sometimes gave me friendly advice, helped me to overcome my hesitation. I dare not name names; for the conception presented here is done at my own risk, including its errors, and no one to whom I owe thanks should be burdened with it.

About my sources I will note the following: for the early history of the Station, I used, apart from otherwise available original material, Dohrn's own writings. These could, however, not simply be taken over, because they are often inordinately long, so extensive that the presentation of 4 years requires almost as much space as this entire book. People and situations are blurred by the passing of the years. No doubt, Dohrn himself would have abbreviated these notes before printing. For important anecdotes Dohrn's words were taken over literally. The basic aim was to let him speak as often as possible through his letters and programmatic essays.

It is easy to understand that a biographical study of his father, Carl August Dohrn, should precede that of Anton Dohrn's own life, not only on account of the fateful importance his father had for the goal and the struggles of his son's life, but also because he himself was an important and very eccentric character in his own right.

Ample material was taken from the Dohrn family archive in Naples, which was thoroughly investigated in many weeks of labor. I was allowed to investigate the material of the Office of Foreign Affairs, the Reich's Ministry for Science, Instruction, and Popular Education, the Prussian Academy of Science, the home archive of Brandenburg-Prussia, the Administration of the German Reichstag, and the Friedrich Schiller University in Jena. Of course, I also read the Dohrn material preserved in the Stettin city library. The Ernst Haeckel Museum offered Dohrn's letters to Haeckel, and Dr. Walter von Keudell in Hohen-Lübbichow let me see the letters Dohrn addressed to his father and to him. I herewith thank the offices and the individuals who played a friendly, helpful part in the development of this work.

It was at times doubtful whether our own time, during which this book was written, was favorable for such subject matter. The present author found it sometimes strange to study this material during the weeks when world historical decisions were being formed. But the execution of the plan finds its justification in the timeless and lasting achievement of the man for whom this book was written: he is one of the greatest and most vital witnesses to the worldwide influence of the German »Geist«.

Berlin-Lichterfelde, August 1940 THEODOR HEUSS

PREFACE · 1948

Since this preface was written 8 years ago, the world has changed. The biography of Anton Dohrn met a friendly reception when it was first published: it was sold out within a few months. But the publisher, pressed hard by lack of paper, could not make up his mind to republish it. The requests for the book continued and had to remain unfulfilled.

Hermann Leins of the Rainer Wunderlich publishing house suggested to me, even while the war was still going on, that I republish the book when the conflict was over. I agreed. The book first appeared during the war, when Germany was closed to the world. The man whom it represents, however, and his work belong to the world. This justifies a new edition. On the whole it could remain unchanged. To be sure, some new material showed up after its first publication. It gave a stronger profile to the part played by Heinrich Dohrn the elder in the life of Stettin. There were also letters from Anton Dohrn's brother Heinrich, and it seemed sensible to add this material in a new edition. But this had become impossible. Where did the city archives end up in the storm of the war? Perhaps they still exist, but they certainly cannot be found today.

For Anton Dohrn himself no new historical sources have been found. How his personality is reflected in the history of science and in some personal memoirs is most impressively expressed in the special Dohrn edition of the Berlin periodical *Die Naturwissenschaften,* edited for his 100th birthday.

The reader who has followed the fate of the man, his family, and his work, and who is familiar with the places of his life and work may justifiably claim the right to know how their history continued. How many friends in Germany, nay, in the entire world, were worried when the reports came of air raids on Naples, and later of ground fighting! Those were months when, in addition to personal worries and fears we quietly thought of the Naples friends, the Station, Marées' frescoes. How great was our happiness on learning that the Station had come through the dangers safely, how great the pain that the Casa Dohrn in Via Crispi, so rich in memories, had been destroyed. Margret Boveri permitted me to add to this new edition her lovely obituary on the destroyed house, which was published in the *Frankfurter Zeitung* in August 1943.

As far as it was possible to form a picture from letters and journalistic documents, the most important ones were added to postscripts in order to chronicle the events. It is incomplete, but it contains, aside from the tragedy, optimistic elements. Is it not symbolic that, amid the clash of arms, when passions obstructed a clear view, voices and actions of helpful care came from neutral and from warring countries? The Zoological Station in Naples was an island, for which the seemingly lost law of international fairness was still valid. That is a comfort, and presents perhaps the most elevating and far-reaching effect of Anton Dohrn's humanity.

Stuttgart, August 1948 THEODOR HEUSS

PREFACE · 1962

Any author is, of course, happy when a book he wrote over two decades ago and whose second edition has been unavailable for some time can be printed again. Nothing had to be changed in the text. The war adventures have been briefly sketched in the second edition. But this preliminary remark needs some explanations. Alfred Kühn, a friend of mine from our days together in Berlin (he is now in Tübingen), felt stimulated by my biography, as he wrote to me, to start a complementary work: *Anton Dohrn und die Zoologie seiner Zeit* [Anton Dohrn and the zoology of his time]. That was welcome to me, being a scientific layman, because I had tried in my presentation not to transgress the limits imposed upon me but rather to aim at Dohrn's humanity. Kühn's great essay, which considers the contemporary controversies and the effect of the Zoological Station on the development of zoology was published in 1950 as a special volume of the *Pubblicazioni della Stazione Zoologica di Napoli.*

My biography, slightly shortened in those parts which were too foreign to an Italian reader, was translated by Cordelia Gundolf and published in 1959 by the publishing house Edizioni Casini. In the meantime, the Station has been enlarged by the building of a long-needed new library. The large collection, unique in the international science of zoology, had been saved by being transported to a small town. The beautiful room adorned by Hans von Marées was thus made free. The development of the murals is, of course, reported in my biography. Now, a special publication, technically perfect with color plates, has been published: *Hans von Marées, die Fresken von Neapel,* Prestel Verlag, München, 1958. Bernhard Degenhart wrote the text, which describes them in a careful, art-historical manner. Long before I became familiar with Marées' most complex work, namely the Naples paintings, I had dared in

1909 to write an analysis of this master's work. I was now asked to write a preface to Degenhart's book. I am glad that, aside from the biography of Anton Dohrn, this beautiful book has also become a document to his stimulating power.

Stuttgart, January 1962 THEODOR HEUSS

I THE ANCESTORS

1 HEINRICH DOHRN

The Dohrn family originates in the Prussian province of Pomerania.[1] There is also a somewhat vague family tradition regarding an immigration from Holstein* and, in addition, a somewhat romantic legend about Anton Dohrn's great-grandfather, who is said to have been an aristocratic Dutch student who fled because of his involvement in duels. The early dates are more trustworthy; there is no lack of later ›romantic‹ additions.

In the county seat of Barth we find a dyer, Johann Melchior Dohrn, born in 1688; his place of birth is unknown. In 1713 he married Blandine Herchert, who came from a Barth family that sat on the town council. One of their eight children, Friedrich, born September 13, 1719, became a surgeon. His fate took him across the border to Poland where, as a Protestant, he appears in the Association of the Officials of the Bishop in the province of Posen. The register in Greifswald** lists him on December 19, 1754 as »*Bardensis* (stemming from Barth) and *Principis Theodori Czartoryski Episkopi Posnansiensi et Warsaviensis chirurgus.*« This means that, ambitious and industrious, he rose at the age of 35 from the trade of barber to the profession of medical doctor. There is no report on his character. When Posen became Prussian during the first partition of Poland, Friedrich Dohrn was taken over as a Prussian governmental doctor. His marriage to Anna Burghardt produced four sons; two of them started a family line in Posen and West Prussia and a third, Heinrich, born in Posen on December 6, 1769, returned in 1796 to his father's home in Pomerania. The father died in Posen.

* A Prussian province.

** A town in Pomerania.

Stettin, where the young man settled and immediately received membership in the merchant guild, was at that time the most important port of the Prussian monarchy, all the more so since it had expanded toward the east in the preceding years, an expansion legitimated during the peace of Basle. The young firm of Jahn and Dohrn imported colonial goods and tropical fruit, dealt in wine, took over insurances, produced soap. It grew quickly in all respects and the immigrant soon achieved an honored position. This is confirmed by the fact that in 1809 he entered the city council without pay as town councillor, after Baron vom Stein * had passed a bill for a new type of government for cities in November, 1808. This public activity lasted only 3 years. Heinrich Dohrn was not a politician with local political ambition, but rather a man with a sense of duty. The time when he was working for the town hall in addition to this own business was a most difficult period for Stettin: the fortified city had been in the hands of the French since 1806, and the well-ordered pace of business had given way to uncertainty about the future. Taxes and confiscations pressed on the community and its inhabitants, old firms broke down, the continental embargo paralyzed shipping and trade. Clever adaptation and skilful risks were needed to maintain assets during these times; in this Heinrich Dohrn was successful. He had bought property in Burow, outside the town, which he later sold in 1817 to acquire the splendid vicarage of Hökendorf, close to Stettin, as his summer seat.

In the year 1817 he was in the front line of commerce in Stettin. The »French period« had ended on December 5, 1813, but it had caused much misery and privation. The siege of the fortress had drawn out over months, and even after the battle of Leipzig ** the

* Karl Freiherr vom Stein (1757–1831), Secretary of the budget in Prussia; worked on many reforms during the Napoleonic wars, mainly on raising the bourgeois class to the standards of the aristocracy; education was his main means to this end by changing the school system.

** This was the famous battle in which Napoleon, on his return from Russia, was defeated by Prussia.

French occupation army hoped for some possible change of affairs. Finally the Prussian army had to attack the important city even while trying to protect it. Only after peace was certain could it become clear whether the enterprise of the citizens was still alive. It was a period when the exhaustion caused by foreign politics was enhanced by inner political disappointments. The bourgeoisie, called upon for collaboration and independence, at the same time had to work with an active high-level bureaucracy of Stein's school to begin its new tasks. Although in 1816 Stettin could not reach its goal of becoming a tax-free port, it had great advantages. As a result of the new Prussian trade laws of 1818 (no more customs charges from their own country and simplification of the tariffs of foreign countries) the commercial forces were liberated and in turn motivated industry.

Against this historical background we see the man who, by founding Pomerania's Provincial Sugar Refinery [PPZ] in 1817[2], created the first larger industrial enterprise in Stettin. Earlier attempts of a similar kind had not succeeded; now he made a new attempt, retaining whatever was still useful along with existing personal experience, and brought together a circle of wealthy men – certainly just one incident in the history of local industry which was followed by hundreds, later thousands of parallel cases in the period of early capitalism. A new wealth developed, a bourgeois wealth; in many German cities a new kind of patrician family developed during these decades, whose stature was expected to prove and confirm itself by promoting a sense of the public good and by supporting the intellectual power of a wider awareness. In this sense Heinrich Dohrn, together with the shipowner Fr. W. Gribel, was not only the creator of the first Stettin industry, but also the founder of a family tradition.

In 1817 he was appointed first Director and confirmed as such in 1820 after the final settlement. He led the sugar refinery resolutely and skillfully through initial dangers. The »boiling house« refined cane sugar from overseas, mainly from the West Indies; it was brought by boat directly to the factory after a new building had been finished in 1824. With a courageous view of the future, he had secured a large tract of land early on, and the development proved the optimism of the founder to be justified: the Stettin factory rapidly conquered the eastern markets of Prussia and business expanded quickly. Of course, there were occasional setbacks, but in a few decades the starting capital of 200 000 Thaler was multiplied by surplus and savings. Some critical problems then arose in the Stettin factory, namely, how it could and should deal with the rising production of beet sugar. The latter had some tax benefits, whereas the general shipping interests of Stettin tended toward cane sugar. Heinrich Dohrn initiated these considerations, carefully observing the new development, but he did not carry them out further. He remained first director until his death on May 18, 1852. Since 1845, however, he had been represented by his son, Carl August, at first as substitute and after 1847 as adjunct. Residing in Hökendorf, the old man was consulted, and lived to see how, starting in 1851, the factory changed over almost completely to the refining of beet sugar.

A very characteristic portrait of the old man exists: the embroidered cap sitting somewhat crookedly on his head appears somewhat old-fashioned but comfortable. The expression on the distinguished face is firm and powerful, its features strong and large, the eyes free and clear, a certain harsh sternness about the mouth: an intelligent man of strong will and reserved dignity. But this is not his only feature. He was also an eager music lover, and he played the violin in a quartet which met regularly at his house. Their playing is praised along with their taste: familiar with the classical masters, they tried to interest the son of the house in their activities. Dohrn's house was a center for music. The composer Karl Loewe worked in Stettin from the 1820's to the 1860's as musical director and cultivated public interest in this art so much that Dohrn's son Carl August, who had traveled widely, considered his home town to be the most musically educated city of the entire world. [3]

2 CARL AUGUST DOHRN

In 1810, Heinrich Dohrn had married the widow Louise Toussaint, née Piernay. The marriage remained childless, but a son had been born to him from a free union with Johanna Hüttern on June 27, 1806. The mother, only 24 years old, died in July, 1807 from a »nervous fever.« Little information can be found about her. She was a soldier's child, »daughter of Johann Hüttern of the Wrangel company in the Regiment of Queis, and of Sibilla Blauen, daughter of the former soldier Christof Blauen of the same regiment.« On March 3, 1814, the son was made legitimate.

Carl August Dohrn was an odd character in every way, a late child of Romanticism with genial and eccentric features, talented in many ways, and full of potential for great accomplishments, which remained, however, promises rather than fullfillment; he was a dilettante, but in a grand style. In his youth he captivated people by his great social talents and his famous singing. Again and again he was seized by an urge for traveling; he lived for and through human relationships, and was an untiring, even insatiable, correspondent until a very old age. In between he became homebound, silent, shut up within himself, even ill-humored; he withdrew from his environment and hid behind his passion for collecting and sorting insects, a passion which did not lack a certain pedantry and which had seized him in his mature years. The order and precision of his insect collection formed the basis of his scientific reputation. This particular aspect of his various characteristics was to have an important influence on his youngest son's choice of profession. But Carl August also remained a fateful figure in Anton Dohrn's life in a more general way; their relationship continued to be a strong and occasionally very painful interaction of competing forces.

The tension between father and son, often close to a complete break, was repeated in both generations although for different reasons. Carl August, in his older years, would fight Anton's independent decisions, forgetting that his own development had been achieved by rebellion. As a rebel son himself, however, Carl August was victorious, because his father submitted to his will and his stepmother assumed a mitigating role. She had taken over the education of the teenage boy out of a loving sense of duty. In a biographical study which his son, Heinrich, wrote from tales and letters after his father's death in 1892, an approximate picture of the situation emerges: the young Carl August had been left much to himself, living alone in the city house during the school months. Undisturbed, he could develop his urge for a free and independent mind, but at the same time his father wanted to shape him through strict discipline – something difficult to achieve considering the difference in temperament.

Pictures from his youth show an odd mixture of dreaminess and defiance. Later, he gave an account of this to himself and to his father: the more he felt that he was expected to do something by force or direct order, the more rebellious he became. Although he had, in fact, inherited a musical talent even greater than his father's, he refused to follow his father's model of carefully handling chamber music with a string quartet, etc. His natural talent was for singing. His sure intonation caused his superiors of the choir of St. Jacobi to let him sing solo parts; and for the sake of singing he eagerly and successfully learned to play the piano. That led him beyond chamber pieces to operas and oratoria. Aside from his musical activity, he did a lot of somewhat wild and unselective reading, eagerly swallowing all the literature he could get hold of, both German and foreign. Goethe and Shakespeare were his earlier gods. It is difficult to say whether he turned to the Romance languages and intellectual world at this early stage. His first knowledge of French was acquired from the soldiers of a guard unit which was stationed near his father's house.

His educational principle, which wavered between narrow discipline and trusting liberality, was a source of worry for his father. But there was no reason to worry about school. The quick understanding of the young Carl August helped him to advance quickly.

He was not yet seventeen years old when he passed his final exams (Abitur) at the Marienstift high school (Gymnasium).*

The final school report declares that his natural talents would have let him achieve satisfying progress even in mathematics and history, had he not lacked a more permanent interest. His knowledge of languages is praised, particularly his gifted fluency in French. The teachers were satisfied with his »quiet manners and his love of orderliness . . . through his cheerfulness and friendly helpfulness he won the affection of his fellow students.« He started his university study in the summer semester of 1823 in Berlin. It is striking that the young man, who later traveled so much, attended only this one university until 1826.** He registered in the law school – a compromise solution, without doubt. Quite understandably, the young student was not clear about his future profession. The father may have thought that, by studying law, his son's lively mind would find the discipline which would make him capable of finding a lifetime profession. Carl August did not enjoy this field, and in his later life made little use of the knowledge he had acquired. Even in his correspondence, which is rich in allusions and excursions, or in occasional publications, legal references rarely occur. We know little of the scientific yield of that period. Except for a lecture by Ritter on logic he took courses exclusively in his field, mostly by Savigny and Raumer. Neither philology nor natural science interested him. At the appropriate time he took his final examination. Participation in the literary and musical life in Berlin, so far as he had access to it, was more important for the rhythm of his life. At that time he started his friendship with the musicians Felix Mendelssohn and Heinrich Dorn, who was later musical director of the Berlin opera for several decades.[4] In his *Memoirs*

---

* The Gymnasium is a special high school for future university students. Its final examination, called *Abitur*, serves as entrance examination to a university.

** Most German students attend more than one university of their choice, until they settle down to write a dissertation.

Dorn related how in the winter semester of 1823–24 two people sat opposite each other in a bar, »one of whom commanded the attention of the other by the Napoleonic cut of his face, but even more so by his strikingly restless behavior.« The observer watched how the other »plays unsuccessful Japanese knife games, throws bread balls to the ceiling, emits a medley of whistling without measure or rhythm, but uses this natural instrument with great virtuosity.« They started talking about a problem of rhythm; that evening they played the Finale of Don Juan four-handed without music; and they proved to each other their mutual admiration and became »inseparable friends.« A third person then joined them, Constantin Holland. Dohrn had made this stranger angry in a café by whistling the Papageno motif after each missed billiard stroke. An exchange of cards had followed and then a duel in which no blood was shed, with the duelers then becoming friends. Dorn tells how difficult it was to take Dohrn into good society, open to talented young musicians, because he refused to wear tails. But he was always good company during the pleasant hours of comradeship. These recollected anecdotes have the serene charm of an unregulated life. This was probably the time when Carl August, in his letters to Mendelssohn, signed as Spihna, adding the »h« to the Latin word for thorn (in German: Dorn).

In his years as a young barrister in Stettin he retained this lively élan. The young jurist became attached to Karl Loewe, whom he called the »fat darling of Apollo.« He sang and reveled, putting off the decision about a profession. Although he could not conceive of himself in a solid law firm, he felt on the other hand the limits of this artistic strength, realizing its receptive and at best reproductive character.

In the winter of 1830–31, at the age of 24, he met a woman, and the subsequent fight for her gave a decisive turn to his life, which for years became almost adventurous. She was married to a Stettin merchant, Weiss, from the Berlin family Dietrich, with a strong musical tradition. On her mother's side she was the granddaughter of the Mainz bassoonist Johann Nepomuk Hamel, praised in musical history as a virtuoso on his instrument. The woman was willing to divorce her husband and follow her passionate lover: the divorce was completed in July, 1831. But the planned union encountered the strongest opposition from Dohrn's parents, who were upset over their son's having broken up the marriage of friends. And how could a young, immature man who had no bourgeois position take upon himself the responsibility for another human being? The refusal of his father was unconditional.

But the son did not give in. He left Stettin with the vague intention of creating some kind of position for himself, somewhere and somehow, with the help of his connections and his talents. The beloved woman, who bore him a daughter in August, returned to Berlin and courageously began to build a modest life for herself by teaching the piano. She was waiting until things improved – she was ready to trust the chivalrous loyalty of the man.

What Carl August Dohrn really intended to do remains vague. He spent some time in Munich, then went to Paris. On his trip there, which went through Weimar, curiosity as well as the encouragement of the doorkeeper of the Frauenplatz (where Goethe lived) caused him to visit Goethe. Since he could tell him something of the latest works by Mendelssohn he later considered this meeting, which he dared to wrench from fate, not to be too immodest. In Paris, he tried to enlist in Brazil for the defense of the royal rights of Dom Pedro. Did he seriously mean it? Although his personal needs were moderate, life had up to then spoiled him and kept him from the worries of earning a living. Could he count on his father's anger disappearing? In fact, his father was deeply alarmed by his son's departure. Although he opposed the planned marriage, he did not want to lose his only child completely. Through friends he saw to it that an exchange of letters was begun, and that the son, without realizing the source, received the means of subsistence. It was an uncertain situation which permitted the son to search and plan without suffering great poverty. To know that his son was in need and to leave him in that situation would have been unbearable, both humanly and socially, for a man whose financial situation was improving very pleasantly during those years.

Such was the situation when Alexander von Humboldt,* like fate itself, intervened in the relationship between father and son. Felix Mendelssohn, then at the height of his youthful fame, was able to arouse some interest in his friend and his friend's circumstances in the mind of this great scientist, who was always open to human problems. Humboldt offered his help as a counselor and mediator. Carl August could not have wished for a more effective spokesman. The Chamberlain of the King, the man whose scientific name was known over the continents, wrote letters to the Stettin merchant about his son and promised to influence his future development. This forced the parents to change their position. In fact, a mutual approach and reconciliation occurred, although there was no change in attitude toward the marriage. Carl August had once told his benefactor that the would like to become a merchant; Humboldt took the idea up, persuaded him of its soundness, and suggested this plan to his father, who was very satisfied with it, seeing in it his son's turn toward a more realistic concept of life. He was certain to ease his son's way toward receiving a thorough education. That it could not take place in Stettin was obvious after all that had preceded. Perhaps it would be best to acquire such an education in a foreign country.

The relationship with Humboldt started in 1832 and remained active until the latter's death in 1859. In Humboldt's later years, it was not only pleasant but useful to Carl August's career. For him, Humboldt's existence remained the measure of human greatness. He felt confirmed by it in two ways. First, when he turned his literary interest toward the Spanish cultural heritage to which Humboldt was so lovingly attached. The second was his turning

---

* Alexander von Humboldt (1769–1859), scientist, traveled to South America (1799–1804); wrote his main work »Cosmos, a World Physics,« which included botany, zoology, mineralology, etc. He also wrote *Voyage aux régions équinoxiales du Nouveau Continent.*

toward scientific endeavours. Dohrn devoted his later decades almost exclusively to their energetic pursuit and broad support; in this, he felt that he was the master's immediate successor. No one had taught the Germans the way Humboldt had to appreciate, with an inquiring and gratifying devotion, the realities of the life of nature and to comprehend the minute and the inconspicuous within the context of the great natural laws. Humboldt's letters were preserved as precious objects. Dohrn never forgot what this man once meant to his personal life. In an essay, written for Humboldt's 100th birthday (1869), Dohrn recounted how their relationship became more intimate: a song he once brought home from Almeria* moved the old man. Decades before, Humboldt had heard it sung by herdsmen in the Andes and it awakened memories in him. From then on, Dohrn did not have to be announced formally to the extremely busy man; he was admitted freely and without ceremony as a friend. In a cheerful mood, Dohrn tells how he won from his friend, a brisk and voluble speaker himself, the freedom of continuing his own speech without interruption. Their intimacy, however, did not diminish Dohrn's reverence. He still rushed to the side of his dying friend.

Humboldt had suggested to him to continue with his Spanish work and to leave the entomological matters to others – because he felt that Dohrn was the only advocate for Spanish literature. Nevertheless, he drew Dohrn eagerly to his own studies on the relationship between plants and the spread of insects. Carl August was proud that he could assist him in this field.

Humboldt's advice to become a merchant – if it was ever more than just a diplomatic suggestion for a temporary solution to the family's discord – did not do much credit to Humboldt's knowledge of human relations. Carl August Dohrn, with all his talents, became no more like a »real« merchant, than he had been a lawyer; he lacked all business sense. But at least the life style of the years that followed had acquired a symbol; it referred to a marvelously

* Spanish province.

free life of travel in which he was willing to learn, and did learn much. Of course it also contained office experience, accounts, bills of lading, drafts. These business items remained an embarrassment; they were begun without pleasure, executed with a sense of filial duty, and repeatedly interrupted. Letters from Stettin gave friendly advice. The father wanted to know how his son enjoyed mercantile activity, whether he liked it, what progress he made. But he also urged him to learn languages, to get to know the countries, their peoples and their arts, and not to neglect his music. No doubt these encouragements found a very willing reception and a child-like obedience.

Carl August Dohrn called the period from 1831 to 1837 six immoderate, restless years of wandering which followed the six moderately settled years of learning.* He apparently visited home only once, for a short time; during that period traveling was still a cumbersome matter. (It is somewhat confusing and perhaps unnecessary to follow the individual stations chronologically.) Before beginning his mercantile education in Hamburg in 1832, Dohrn undertook his first trip from Paris to Norway and Sweden. He had little liking for the way business was managed in Hamburg. He wanted to observe the work of a merchant on a full-time basis, like any other occupation in the state, »but I doubt that I will ever develop a love for it. Actually, to put it bluntly, I am old enough to discover the conviction, in everything I do and love, that I have no sense of money and numbers. Personally, matters are even worse . . .« The conclusion of such thoughts was: »I answer this modestly and calmly: God and you may know . . . It does not seem unnatural to me that I share with you, as your child, the wealth which you earned through diligence and good luck. I am very grateful to you for this and for everything you made me learn. We have different views about what it means ›to be useful to the world‹. For you, an activity, as you practiced it, was both a necessity and a pleasure; it was also crowned by success. And nothing is

* The titles of the two parts of Goethe's *Wilhelm Meister*.

more natural than your idea to see in an activity the main purpose of life ... I have little or no sense for acquiring wealth, but I may rightfully reject the idea that I am wasting it since I am satisfied with anything. Still, I can enjoy good things gratefully,« (November 12, 1832). These reflections are quite characteristic of the open manner in which he spoke to his father. They created »very ill humor« in the father and, in order not to endanger their relationship, his father did not answer them. But the theme is continued in other letters. Carl August rejected his father's suspicion of laziness, finding that: »two thirds of the very active people here are really destroyers of time before God.« »I find it impossible not to be active – my character is not phlegmatic enough for it. But you may blame me for being indolent since you want to measure my activity according to its present result which is, of course, zero. Give me time – it will happen.«

During the winter in Hamburg, Carl August became unwell. He was supposed to and wanted to go South, suggesting that his father let him take a trip to the West Indies and to South America, since that could perhaps help the business of sugar refining. (The fact that he had learned Spanish from a former Spanish officer, who had been driven by fate to the Elbe, was the most productive result of that year.) The phrase »sugar refining« had been well chosen. The old man was willing to take up any remark which showed something »mercantile«. He agreed at that moment not to Brazil, but to Marseilles. There the young man could observe a new method of refining sugar and report back on it. Carl August agreed: It should, however, not be taken too literally or connected with pedantic accounts – the first step, after all, is taken. He did, in fact, take care of the things his father expected; he presented the Marseilles methods of fabrication, ending with the following words: »In writing this, I could use neither the ten commandments, nor the twelve tablets,* nor the principles of counterpoint, nor the

* The first Roman Code of Law was written on twelve bronze tablets in 450 B.C. by the Decemvirate (10 selected men).

profile of the Venus de Medici – therefore, my writing will probably be incomplete and biased.«

Marseilles was partly an interlude, partly a pretense. His father, becoming resigned to the fact that the mercantile trip had turned into an educational one, did not spare advice or suggestions. Thus, a plan emerged to spend a year in Italy and perhaps to visit Spain. North America was being discussed, but the son suggested taking that route by way of Brazil. He obtained his father's permission: in Brazil one could study the plantations of cane sugar and it could be useful to know them if, one day, he should help his father with his sugar production. In that event, he would not only not be patronized by the father, but would be strengthened by his own knowledge. All this proved satisfactory. Humboldt helped to acquire the necessary passports. He, as well as his father, warned the enterprising young man not to enter politics in the countries he wanted to visit – there was a lot of unrest in the world during the 1830's. The son answered that there was no need to worry: »My individual view of the great questions of our restless era is anything but absolutistic, but I feel even less attracted to playing the preacher on the roof and I have a great distaste for all conspiratorial connections.« »I will observe and learn only, not teach or give improvised roles as a guest-speaker.« What attracted and occupied his mind was what sensuously and immediately fascinated him: »Climate, vegetation, and language are the true borders of peoples, rather than the things the gentlemen of the pen or the sword prefer.«

A year in Italy, some months in Spain and Algiers, a visit to London, the trip to Brazil, where he stayed more briefly than planned – when he spoke of these trips later, he reminded the younger generation that there were still no railroads at that time. Prince Pückler, the great garden architect of Muskau, whose travel descriptions were widely read, traveled at that time in a similar way. It is almost symbolic that Dohrn, who was in Algiers, met frequently with his famous colleague. Dohrn appeared in friendly remarks in Pückler's *Semilasso's Penultimate World Trip*. The parents in Stettin felt a little flattered by this – friends pointed it out. Carl

August did not particularly like the Prince, who would have liked to take him along to Egypt as a kind of secretary.

For a while, Dohrn himself considered describing his criss-cross trips, but this did not materialize. The diary entries and the letters to his parents are rich enough in observations but, at the same time, they are full of personal remarks which lack an inner freedom and objective distance. Without seeking anything abroad, he found a great deal. In a letter from Spain, he called himself: »the vagabond, happy in God and the world.« A gaily conversational tone prevails in the letters, which were meant to give his parents pleasure. But there are also serious remarks on poverty, want, clericalism, etc. He played the romanticist who wanted to absorb customs, colors, culture, song, odor, but he also approached the generation of »Young Germany«*; such critical reflection threatened – but did not destroy – the observer's vision and detachment.

As an old man, Dohrn sometimes dug out memories from his journal, conjured up persons he had met and with whom he had had a friendly exchange. The time spent in Rome was particularly rich. He met Thorwaldsen,** the landscape painter Catel, and Horace Vernet, who created great pictures of French history. His knowledge of languages made it easy for him to initiate contacts everywhere. People liked his social talents and the easy-going energy, which made him always ready to sing. Perhaps the permanent profit of his travels was the fact that he brought home an immense amount of foreign popular songs. He did not treat them from a philological or folkloric point of view, nor did he collect them systematically; he just picked them up on the streets. When a tune touched him, he had the whole song sung to him again, noted, with a sharp ear, characteristic modulations, and in the South he watched the gestures of the singers. He quickly wrote down the

---

* »Young Germany« [Junges Deutschland], a partly political, but mainly literary movement, which opposed Romanticism and was interested in politics and social problems.

** Bertel Thorwaldsen (1770–1844), Danish sculptor.

notes, added an accompaniment, and had his own reproduction critically examined. His very reliable ear and marvelous memory, which had always enabled him to repeat entire operas with his own accompaniment, created for him a treasure that he loved to squander. This knowledge of foreign folk songs, nordic as well as southern, raised his artistic talents above others and created his particular fame.

Humboldt introduced him to the Royal Family. As a successor of Dohrn, Count Schack [5], himself familiar with and loving Spanish literature, honored his efforts: »Until many years later, Dohrn retained the power and bloom of his voice, with which he had enchanted the Prussian Court during the life of Friedrich Wilhelm IV. The King enjoyed this rare man's art of singing, as well as the originality of Dohrn's personality. He insisted that Dohrn appear, at the musical evenings given by the King, in his student coat and cap (which he wore to old age), while sitting at a piano in the castles of Charlottenburg and Berlin. There are many anecdotes of those evenings. Once, in Sanssouci on August 2, 1843, when Dohrn had sung a Scottish soldier's song, the King said: »Hearing this air, one would like to strike them down immediately.« Two months later, Dohrn sent him a »Prussian war song« he had written with that melody, »with the kind wish that your Majesty, during a long reign of peace and blessings, may never find cause to use it for any but an esthetic purpose.« [6] He was also able to use the benevolence of the King for his scientific endeavors.

In the spring of 1837, Carl August returned home after tiring of America and after spending some months in Marseilles and Rome. His father realized that the »mercantile« result of these years of traveling was not quite sufficient for the bourgeois security which he had demanded for a marriage. Therefore, he offered his son a sum which gave him freedom to embark upon his literary and artistic plans. The rest would take care of itself. Above all, he withdrew his objection to the marriage with Adelheid Dietrich and the wedding took place in Berlin on August 8, 1837. The couple settled in Braunschweig, but moved to Stettin in 1838; the confusions of the winter of 1830–1831 were forgotten. The well-traveled man could return to his old home as a different person. In quick succession, three sons were born; Heinrich in 1838, Wilhelm in 1839, and Felix Anton on December 29, 1840.

The first years of the new life were devoted to calm literary studies, which the father observed with benevolent and encouraging patience. He had every hope that bourgeois life would draw the son into its aura of duty. The relationship between the two had evened out. Carl August Dohrn stepped up to his old father's side, first as »substitute« in 1845, and then as »adjunct« in 1847, with the title of director without salary. The business situation demanded the performance of many tasks which, with his quick mind, he easily assumed. The controversy between cane sugar and beet sugar entered its last stage; it was essentially a fight about taxes. Soon it became clear that the »Indian« refiners were on the defensive, then finally in retreat against the agrarian interests. A »refiner group« was formed, which in the spring of 1849 sent Carl August Dohrn as their representative to the National Assembly in Frankfurt. In October of that year, he negotiated, in Berlin, with the ministry and the most important members of the chambers; the confidence of his bearing, his engaging behavior, and also his knowledge of the soil were in his favor – but not much was achieved. Still, his eagerness was a source of satisfaction to his father. His son would be his successor. Indeed, when Heinrich Dohrn died at the age of 83, in 1852, Carl August became the first director of the »PPZ.« He remained in this position for 20 years, resigning after a serious illness, in 1872.

Even during this period, however, his interest was not directed toward »mercantilism.« His very talkative letters rarely mention these things and when they do there is an undertone of self-irony. The presence of expert co-directors made this attitude possible. In the mid-1850's Dohrn withdrew from current business. It was enough if he played a guest role in the offices once a week. The father had made the business great; the son did not endanger the growth, but his ambition and his talents sought different activities.

What they found was in a very different direction. The paradox of the man whose true element was music showed itself in the combined role of translator and collector of entomological material.* From 1841–1844, four volumes of *Spanish Plays* translated by Dohrn were published. The strong contemporary interest in Spanish literature, originated by Schlegel's** translations from Calderón and Tieck's *Don Quijote*, was already ebbing. In the dramatic field, Calderón had been so dominant that the large number of other poets, even Lope de Vega, were barely considered. Some plays had conquered the German stage and remained there, particularly if they were »revised«. For although the Spanish theater had been praised for being full of stimulating material and inexhaustible inventions of interesting situations, it was not easily accessible. Thus, free revisions developed early on.

Did Carl August Dohrn hope to steer the minds of theater directors back to the Spaniards? Later, he spoke with a certain resignation of his attempts which, although meeting a friendly reception by the connoisseurs, did not find a larger echo. Only *Don Gil in the Green Pants* was produced on the stage in his translation. This was the lighthearted story of confusion and disguise by Tirso de Molinas. It was produced in Berlin in the 1890's, but its discoverer and translator did not live to witness its reception. The translations are marked by a great conscientiousness, by means of which Dohrn wanted to come close to the different metrics of the Spanish plays. He was very conscious of the difficulties because German and Spanish have such a different spirit and rhythm, and he saw in this endeavor only the possibility of an approximation. More could not be accomplished. But he wished that in the translation one could sense the original language. In all the plays he made known

* Entomology is the science of insects.

** August Wilhelm Schlegel (1767–1845), the older of two brothers who, together with some friends, created and discussed the idea and practice of Romanticism. He translated Shakespeare and many Spanish plays into German.

in Germany, the musicality of the Spanish language should be heard. He was not addressing himself to readers from the masses, but »to the small circle possessing a sense of enjoyment for a comparative esthetic anatomy.« His works are full of a formal artistry, which at times has something forced but on the whole has a certain lightness and shows a talented treatment of language. That is particularly true for lighthearted plays, for burlesques, and even farces. When he realized the limits of translation, as in puns, he did not shy away from using substitutes. But, in an added commentary which partly explains the foreign language, he lets the reader witness his difficulties.

When Dohrn started on his Spanish work, A. W. Schlegel was still alive, but their personal relationship was not productive. The young enthusiast remembered, without pleasure, the annoying vanity of the old man. At the same time, but independently, Count Schack was working on his studies which appeared in 1845 under the title *Spanish Theater*. Within this framework, Dohrn's work emerges as an interlude in intellectual history. Decades later, as an old man, he returned to his endeavors and published two plays by Calderón in 1878–79. Yet this was only an episode; his hope to win the Germans over to an understanding of Spanish language and culture had long disappeared.

In the meantime, the interest of this flexible man had taken a new direction; he had involved himself in a subject which almost entirely occupied him during the years and decades which followed. The musician, the translator, the humanist became a zoologist. In the fall of 1837, the Stettin physician Ludwig Ewald Schmidt got together some friends of nature to found an Entomological Society. November 6 is considered the date of its founding. On February 3, 1838 »the independent C. A. Dohren« became a member. The printed agenda introduced the future president with a misprint! After two and a half years the »literary man Dohrn« became Secretary and Second Manager. In 1843, Schmidt died young and Dohrn became his successor as President and publisher of the journal *Entomologische Zeitung*, first printed in 1840.

One could dismiss this association as a minor local event, if it were not for a certain historical charm beyond its biographical interest: it was the first association of German entomologists, and lasted for several decades. It was considered, with local pride, to be the third partner of the scientific societies in Paris and London. Later, when similar societies were founded, the people of Stettin found friendly words for the expansion of scientific interests of the group; but there was also a certain fear that the Stettin enterprise and its publications might be in danger of losing their leading position and becoming despised.

Carl August Dohrn had been Schmidt's schoolmate and was impressed by his precise scientific seriousness. In Dohrn's past life, natural science had not played a major role: his childhood collections, gathered in the vicinity of his father's estate, were quite insignificant. In Algiers, together with a collector (whom Dohrn accompanied on his trips because he liked the man), Dohrn occasionally took along some noteworthy specimen and kept it. He was pleased to present one in Brazil to an acquaintance and accepted from him a few samples in exchange – all this was accidental, unsystematic, and lacked scientific intent. One may even say that the beginning in Stettin was caused by his friendly relationship with Schmidt, rather than by any objective passion. Since he was free of business worries, he was eager to help such considerate people and to secure their enterprise. But when he, still the most unscientific in the group, was offered the presidency (perhaps with some reservations), he put his entire ambition and his rich connections in the service of this new activity.

Humboldt, who had originally hoped that his young friend would remain devoted to his Spanish interests and leave the insects to others, was nevertheless helpful. The court of Prussia had planned to give the young musician a medal. Dohrn had a lively desire to deter this. Although he did not lack vanity, he wished to avoid all medals. Later, having been successful in this, he was full of irony with regard to the honors which were extended from all sides to his son, Anton. In the 1840's, he managed to direct the

King's interest toward offering the Society an annual sum for a number of years. Up to the end of Friedrich Wilhelm IV's life, this was granted; first 500 Thaler, then, starting in 1854, 100 Thaler. With this sum, Dohrn could hire a young scientist and extend the connections of the Society.

When he took over the directorship his main concern was not science but rather the education of the people. »Between absolute ignorance and complete scientific knowledge in the field of entomology there are a sufficient number of levels which anybody can achieve.« He was eager »to place science next to the much venerated humanities as a main source of national education.« »It is a noble and worthy goal to give entomology a special place in general education, not as an idol of nature, but as the precious means of knowing and revering God's wisdom in the wonderful household of nature.« There are later many variations on that theme. »A State does well if it teaches the inhabitants to fill their leisure hours (which even the most industrious need for recreation) pleasurably and without regret.« (1849) In the same year, as well as in later war years, he praised the good luck »to lean, amid the storms of the present time, on the eternal unshakable pillars of the Temple of Mother Isis,«* to find there one's inner balance, »without falling back to the indifferent Philistinism of personal favorite pursuits, unfortunately not rare in Germany.«

Over four decades, until 1887, Carl August Dohrn directed the *Entomologische Zeitung.*[7] Then he gave both it and the Society to his son, Heinrich. It is one of the strangest scientific publications ever published. Before he died, Schmidt had already found very good collaborators. Dohrn enlarged the staff. He succeeded in attracting important men like C. T. von Siebold as collaborators. He expanded relations to foreign countries and himself translated miscellaneous information he found in foreign publications. At the beginning, his own professional work was still relatively modest. Occasionally he wrote reviews; he began to describe and to classify. This was then

* The Egyptian goddess Isis was goddess of Nature.

extended; genera and species were treated systematically. Soon the collecting and exchanging activities were pursued on a large scale, whereby friendly connections were established with all countries. He knew only too well that such ordering, sifting, denominating, and interpreting activities required not only enthusiastic pleasure but also sober pedantry. The latter was not part of his natural predisposition so he forced it on himself; yet, eventually he became tired of the dry tone. Then he would interrupt those valuable, but often boring contributions with jokes, anecdotes, and bold commentaries. Already the 1844 volume began with a poem – this continued over the years. The subscribers probably grew used to – at least, they had to grow used to – the fact that entomology was treated lyrically, with all kinds of poetic measures, and in different keys. There were puns in the style of street-singers, strict, complex sonnets, hexameters, doggerel, folk wisdom, and romances.

Was he trying to tell the »members of the vermin lodge« that he was a poet? His knowledge and idea of the nature of true art were too great to demand such an elevated title for his accidental creations. But he abandoned himself without restraint to that »urge to play,« although some opponents were angry and criticized the tone and the intimate attitude. This enraged and pained him. For example, there were years of complicated, angry, or significant polemics with the Berlin entomologist Kraatz, whom he considered a deserter. His sons were angry about these hostilities and the battle reports which today, in retrospect, seem like dated comedy motifs. But Carl August Dohrn had grown into the role of a prince of entomology, being a stimulator and supporter, scientist, and collector.

Indeed, during four decades, he had become the most important German figure for the cultivation of entomological interests. The travels that took him to St. Petersburg, Moscow, and again and again to Paris and London, were now devoted exclusively to naturalist friends and to comparing, exchanging, and buying. His correspondence went to all parts of the world, not only to scholars, but merchants, officers, and missionaries were also invited and encouraged. He helped and advised beginners and he conducted debates with specialists. He himself had now become one, particularly a specialist of beetles. His collection of paussids* gained him broad fame among experts. (It was bequeathed to the city museum, along with all his other naturalist possessions.) His humorous mockery was transformed into a searching, comparing, and identifying zeal, which constantly wanted to locate and name new species and varieties among the innumerable existing ones. »The mono-grafs [counts] and the mihi-barons provide amazingly bountiful Borisflor. Here and there, a Matador engages in combat for a paper crown of immortality.« But soon he found himself in that heaven of entomological immortality, since friends and scholarly societies added his name to this or that species. He took great pleasure in this.

In the meantime, the impressive set of volumes of the *Linnaea Entomologica*, edited by Dohrn from 1846 to 1866, grew. This was an exact work in which no joking was permitted, but it also caused a problem, since it was often questionable whether this long-range project was feasible. Dohrn enlisted the contributors and took care of the commercial troubles. The work gave an account of the contemporary status of the science; and it found recognition. The English Entomological Society conferred its membership to C. A. Dohrn in 1855, and its honorary membership in 1885. The same award was given to him by the Dutch entomologists in 1875 and by the Swiss entomologists in 1876. The University of Königsberg conferred on him the title of honorary Doctor of Philosophy in 1862, in recognition of his editorship of the *Linnaea*; he did not reject it. The following year was a climax, honoring the achievements of the Society, which was then celebrating its 25th anniversary. The German Society of Naturalists and Physicians, founded by Humboldt, convened in Stettin in 1863. Carl August Dohrn took charge of the local preparations and of its management. It gained a certain historical significance, since the young Ernst Haeckel from Jena delivered there his declaration in favor of Darwin's theory which had been revealed a few years earlier. Carl

* Small animals of the family of beetles.

August, who venerated Linnaeus as a great systematist, was perhaps more disturbed than thrilled by what he heard. Next to him stood his youngest son, who anticipated in this a radical turning point in the problematics and methodology of zoological research.

We may consider this situation for a moment: it was more than a difference of generations. Carl August Dohrn was a figure of the Biedermeier era, also a product of the entire educational breadth of that period; he was a collector, not an investigator, a friend of nature, but not a speculative or empirical interpreter of the essence of nature. Romanticism had formed him musically and artistically, but not philosophically. His religious sense was poorly developed and he was not impressed by the Romantic philosophy of nature. One of his favorite authors was Jean Paul. * He recited, for example, the beginning of the *Flegeljahre* over and over again. An examination of his conversations or his letters reveals imitations of his revered master. He enjoyed every flash of thought or wit, constructed monstrous neologisms, and employed innuendo and self-irony. When he found »entomology« in Calderón, or admired – during his trips through art galleries – a butterfly or a beetle in a Dutch or Italian still-life, he showed a true affinity to Jean Paul, and concentrated lovingly and critically on this painted documentation of love of nature. His equipment consisted of the botanizing box and butterfly net, rather than the microscope or microtome. His intellectual attitude was one of grateful contemplation.

When, after his father's death in 1853, C. A. Dohrn took over the inheritance, he had a stately house built in Lindenstreet. It was intended to house his growing collection. He extended a lively but simple hospitality, as he had done in Hökendorf. Whoever came to Stettin, whether they were musicians, literary men, or scientists, were welcome. The city's location increased his relations to Scandinavia. He had a friendly exchange with the Swedish songwriter,

* Jean Paul (his last name Richter is never used) (1763–1827). He wrote many beautiful and idealistic, but also bizarre novels. He had a great imagination and an unusual sense of humor.

Adolf Fr. Lindblad[8], and his interpreter Jenny Lind. In the 1840's, Dohrn had already published some collections of songs by Lindblad, translating the Swedish texts for voice. He had lively relationships with the artistic circle of Düsseldorf, with C. F. Lessing, and the historical painter Th. Hildebrandt, who was a childhood friend. He bought some of Lessing's paintings. A moving moment in his correspondence with Lessing occurs when he asked him to advise their friend Hildebrand about the choice of his subjects. The great garden architect Peter Josef Lenné, intimate friend of Friedrich Wilhelm IV, belonged to the group of Berlin friends. Lenné was particularly interested in awakening a sensitivity to his noble art among the country estate owners. Lenné created the beautiful park of Hökendorf for his friend.

Since Dohrn had gained a certain popularity, he was elected representative to the Prussian House of Representatives in 1859.[9] The new era had awakened liberal expectations, which Dohrn shared in a general and imprecise way. Politics, in the narrower sense, was not his concern. His candidacy derived from his sense of obligation as a citizen, not from ambition. The letters of that time are not interesting from a political standpoint, even though the period was lively and interesting. They complain about committees, and report on beetles. Even during those weeks in Berlin he did not lay down the scepter of his empire. His personal papers say nothing about his position on questions which would soon determine and change Prussian politics. In the full sessions he spoke only twice – aside from short, insignificant reports concerning petitions. But on both occasions the subjects were close to his heart. In the spring of 1860 citizens of Breslau addressed a petition to the House asking for the unification of Italy. Minister von Schleinitz was angry about it, not wanting at that particular moment a debate on Garibaldi * and the Pontifical State. Dohrn under-

* Giuseppe Garibaldi (1807–1882). Defeated the Kingdoms of Naples and Sicily in the War of Liberation 1860. His visit to England revived the cause of parliamentary reform. Fought for republican France in 1870.

stood this, but referred to his long stay in that country in order to express to Italy »his lively participation on many levels .... It would be a reproach, if we were to keep silent.« The other speech, in which objectivity was surrounded by witticisms, advocated the protection of forests, quoting the opportunities it provided for nesting birds, etc. He also instructed the House about useful insects. At the end of the legislative period, he left and renounced a second candidacy. Parliament remained for him an interlude. But it was to be part of his oldest son's fate: the wavering between success and defeat, the passionate and tough dedication to a life of public activity.

C. A. Dohrn's naive egotism sometimes made daily life within the narrow family circle difficult. As long as the children were small, he was an ambitious father with great affection for them. But the tolerant indifference which his patriarchal nature displayed toward his environment and his servants became narrow-minded when his expectations and demands were confronted by the children's own will. His later life was filled with such harsh conflicts, even alienation. His own fate as a young man was replayed, but he did not always maintain the careful solicitude that his own father had shown. His wife was forced to play the difficult part of a secret mediator. Adelheid Dohrn shared the musical interests of her husband, although less enthusiastically. Entomological activity was foreign to her and left her indifferent. During his trips, Dohrn kept up a lively correspondence with her: he instructed her and told anecdotes, while she reported on the children. Already in her forties she revealed a tone of resignation: »You yourself said that I am only a minor sun.« And with a note of bitterness and irony she said she was, compared to him, »only a minor light.«

Adelheid Dohrn was a beautiful woman, with large, even features. Old age made them sharper, not milder. She took life more seriously than her volatile husband. According to the memory of her son, who revered her, she bore elements of melancholy. He interpreted his own paralyzing depressions, which continually interrupted his active life, as an inheritance from his mother. The entire burden of the household economy rested upon her. She calculated and balanced the accounts, took care of the investments and administrative aspects. Carl August had no sense of money. He knew how to spend it moderately while traveling and collecting. Personally, he was a modest consumer of food, drink, and clothing – he always wore identical suits made of identical material. The tailor had to estimate approximately by eye whether or not the merchandise would fit. The only pleasure that Dohrn was adept at was a devotion to the regular smoking of strong, imported cigars, which he selected according to the time of day. Responsibility for the finances made Adelheid's character even more sober. The innumerable letters she wrote to her children in a large, clear, sharp handwriting (which strangely differed from the small, hurried scribble of »C. A.'s« letters) are full of household instructions, financial information, and business suggestions; all rather worried, and laced with bitter remarks about her husband. But alongside this, there existed a sense of public interest that is more vigorous than her husband's. She eagerly read newspapers, which he only skimmed through; she sympathized with her son's political controversies and meditated upon the parliamentarians, ministers, and election fights. While this is not very original, it is well thought out and shows her definite interest.

On September, 13, 1883, a few days after her 80th birthday, she died in Hökendorf; her husband survived her by nearly 10 years. Dohrn was of an almost indestructible health. In 1872 he had a serious attack in Florence; his son had hurried to him and his life seemed lost. But the sick man recovered and mentioned his newly regained health to the entomologists in a *Conversation with Charon*. »Let me still entomologize for a while before you collect me to the eternal sleepers.« He entomologized for another 20 years. Along these lines, Count Schack tells a story about him. Skating at the age of 82, he broke through the ice, saved himself and had an hour's drive home in his carriage in his wet clothing without even catching cold. He wanted nothing to do with doctors; his interest in matters of modern hygiene was entirely theoretical. He compensated for a life style of defining, classifying, cataloging, and writing about insects by taking demanding hikes during his trips or in the familiar forests of Buchheide. He was always busy and his recreation consisted in varying his occupation.

Carl August Dohrn felt himself to be a child of fortune, but could never rid himself of a secret fear. He was almost grateful when fate, once in a while, would bring him some mishap or disappointment. He did not challenge his fate, but let himself be driven by it with an elated unconcern. The »diis faventibus« * of his letters betokens not only the humanist's pleasure in quoting but the immediate and fundamental feeling of a sensuous, unreflecting man – of an elementary nature. At the age of almost 86, on May 4, 1892, his life, vigorous to the end, ended after a short sickness. [10]

* If the gods are willing.

II YOUTH AND DEVELOPMENT

1 THE SCHOOL YEARS IN STETTIN

Before Anton Dohrn left the gymnasium at the age of 19, he wrote a summary of his life. The document is characteristic not only of Dohrn but of the school as well; he summarized his life up to that point with easy-going cheerfulness. His tone is relaxed and indulgent toward the school: he satirizes his own self-consciousness as he tells about school adventures at various ages, analyzes individual teachers and their good or mediocre success in teaching, talks about his relationship with his siblings and about the spirit of the parental home which »the outstanding men of art and science of the period frequented.« The document conveys the feeling that there was a good relationship between teachers and students in the institution.

Discussing teaching, he says of two professors: »Their classes were more of an exchange, created on the one hand by good will and on the other by respect, as it happens between teacher and student.« His love for some teachers accompanied him throughout his life. The old Pitsch was always visited, and the fortunes of his children remained a friendly concern.

Altogether, the author of this summary was at that point satisfied with the workings of his own fortune. He even thought that occasionally he should have been treated more severely. Out of good intentions one professor had overlooked certain things and advanced him too quickly to the upper classes »which furthered not only my laziness but also my forwardness.« This reference to his laziness seems to be only a manner of speaking, a way of describing his temporary loss of interest. It is hardly thinkable that this boy with his lively interests, who was going to be a man of restless diligence and painful self-control, could have wasted his time.

The parental home was full of children who played musical instruments. Anton, who learned to play the piano, wrote about that passion which would later, although in a mostly passive way, accompany and enrich his entire life. But he says: »Although Felix Mendelssohn was my godfather, I was, among my brothers and sisters, the least endowed with a muscial ear.« Above all, his brother Wilhelm with his eager practicing was superior to him; he lived for music. His oldest son Georg became, after the turn of the century, musical director in Breslau and the leading man in the musical life of Silesia. The inheritance of the rich talent of Carl August also came to life in the descendants of his daughter Anna: Wilhelm Furtwängler was her grandson.

The original artistic talent of Anton Dohrn found its ultimate expression in the creative, organic formation of his life's work and in his superior style of leadership; in the document written for the school, however, he points out that, whereas his brother »devoted himself entirely to the noble art,« he himself was dedicated, »horribile dictu, to a collection of bedbugs.« This is of course also a kind of inheritance from his father, who early on allowed his sons to participate in his entomological work and found willing followers in the eldest son, Heinrich, as well as in the youngest.

When Anton was 19 he wrote for his teachers not so much his speculations on his future, but an actual report of it. He told them that he was a collector of bedbugs: »Or, to use a learned expression, I became a *Hemipterologist.*« At the age of 16 he had published in his father's journal a one-page article, *Hemipterological Thoughts by Anton Dohrn.* Later he wrote that »with this my name was given over to eternity.« He immediately started a large catalog which occupied him for several years. He apologized that his school work suffered somewhat from this, but explained to his teachers the meaning and the difficulties of entomology: »A precise identification requires more sharpness of distinction than the faultless dissection of an elephant.« Continuing in the same good-humored vein, he tells how much he benefited from several long trips on which his father had taken him, to the Rhine and to Düsseldorf with its circle of artists, to Switzerland, and excursions on the Monte Rosa. But the teaching style of his professor in his senior year, who questioned, discussed, and tolerated opposition, renewed his interest in school, »since he was altogether a happy debater.« He found that he was a mixture of »a young man and a boy«, but believes, sincerely, that the teachers will overlook his gaps, »since they know me not only as a student but also as a human being«. This strange work ends with the sentences: »The study I selected holds me, body and soul. It is the sciences, particularly physiology and anatomy. Whatever can be achieved through love of the subject and natural talent, I hope to achieve. The external goal I try to reach is a professorship at a university.«

His report card of March 26, 1860, recognized that the senior »had studied seriously and successfully in a special field of natural science«; his »alienation« from the prescribed fields of studies he had finally overcome through diligence and participation. His grades are all »Satisfactory« – there is a »Good« in physics. »The most important phenomena of nature are well known to him.« More grievous is the case of religion. »His knowledge of Christian doctrine is extremely deficient.« His linguistic abilities are acknowledged, where there are gaps as there are in mathematics, the student knows how to »use skilfully the material known to him.« In the field of German, it is said, »His logical knowledge is mediocre. He handles the German language extremely well, but his manner of thinking is one-sided.« Carl August Dohrn did not worry too much about the school problems of his sons. He made them full partners in his game of L'Hombre* and also taught them other ways of playing cards, but Anton did not consider this a special educational achievement. In retrospect, however, he was grateful to his father for giving him the »intellectual protoplasm« from which his work derived. »I am grateful to my father. He was never angry with me when I did not know at school the Greek verbs ending in ›-*mi*‹ or when I looked on equations with three

* An old card game.

unknowns as a fatal plague; but I was not allowed to ignore quotations from Faust or Henry IV, or not to recognize immediately musical motifs of the Ninth Symphony or the great quartets of Beethoven« (April 14, 1897).

Friendships that might become important in later life did not exist in the youthful circle. The decisive energy with which Anton forced himself to do independent work, the dislike of boyhood or student drinking, the contact with great people from various countries who visited his father, gave him a maturity which, without arrogance, marked a distance from his young comrades. Up to his later years in Jena, where he found an adequate group of friends, he preferred the contact with older persons from whom he could learn and whose debates stimulated him intellectually.

The »scholarly activity of the student« which manifested itself in a few contributions to the *Entomologische Zeitung* and in the cataloguing of his own not insignificant collection, was essentially descriptive, but he showed a critical impartiality in his efforts to show that some species which were designated as different and given different names by different investigators are identical. Later, it became his concern to liberate the subject of entomology from being simply an identification procedure, necessary as that may always be.

His father was often kept in Berlin by his political obligations. The son represented him as a kind of assistant editor, reporting to him about his collecting. Thus, he said, »Murray's shipment is disappointing . . . only one bedbug which, although called *Dohrnii Murr.* (named after your son!) did not answer my expectations.« Aside from entomology, he was greatly attracted to politics as well in the following decade. Carl August had to be instructed by his son that, according to the constitution, the King was not limited in his choice of senators. His brother Heinrich, 3 years his senior, who for a long time adopted the style of a wise superior – until he became a great supporter of his brother's efforts – wrote to their mother on December 30, 1859: »How he (Anton) developed his senseless desire to read Hegel's *Logic* I do not know – the idea is not his. It is lucky that he will understand and digest the nonsense in it (next to some useful thoughts) as little as many sensible people. Despite his philosophic studies he still seems to believe that ›bugology‹ is the center of the world...«

2 STUDIES

Anton spent his first semester in the summer of 1860 in Königsberg; later, he declared that this period was a total loss for him. Already after a few weeks he wrote to his mother that he was yearning for the day he could leave; there is nothing responding to his emotions which »I would even call softness,« although there is no lack of family invitations: »Every philistine invites us for lunch or dinner and in each house we find the same philistines assembled.« One of Carl August's most important collaborators, Hagen, lived in that city. He pleaded that the father should give his son a decent microscope; they worked together. He was introduced to Johann Jacoby, the representative of the East Prussian Radicals. This man's sharp and sarcastic dialectic was not without influence on Dohrn's way of judging the precarious situation of Prussia, which was gradually emerging. University life did not satisfy him at all; he studied histology with Burdach, botany with Caspary; both confirmed that he studied well and with regular diligence; he himself confirmed in his letters their joyless boredom. In ill humor he complained over and over again about the bad pavement – the scope of his experiences seemed rather narrow.

Then he went to Bonn for three semesters. First, however, he served in the regiment of the King's Hussars. His enrollment at the university was only a matter of form; science rested. And even in the fourth semester he had no contacts at the university on the Rhine. But military life was acceptable. In the beginning he was confined to his room for 48 hours because he had exceeded his furlough – his sister, Anna, had been married in Hamm. At first he did not report this mishap; after a few weeks he reported: »My position is quite good.« A soldier's life did not occupy his mind very much: »If there weren't the benefits of riding – for the rest I don't give a dime.« He enjoyed horseback riding: »Mine is the most beautiful horse in the squadron and the best jumper.« He told his mother, not without pride, that »at the visitation of the Commander I was the only volunteer who received special praise . . .«. This was not a momentary courtesy; the commander also inquired about him later on. He found that in the west people were more educated than he had expected – his sergeant owned English and French dictionaries, Walter Scott, and legal works. He did not like the fact that his attendant stood to attention at every word: »I am not made to be a dignitary – I am too polite for that!« He spent the evenings with the newspaper, with the piano, with Goethe – or with a game of L'Hombre. His father may have liked to hear that.

The continuation of Anton Dohrn's military career was less fortunate; it broke off with an anecdote characterizing the period more than the man. The report of the Bonn Commander called him a »very moral and educated young man,« and found his knowledge of military matters sufficient to recommend him as an officer in the reserve. He therefore reported to the first exercise in Demmin at the 2nd Uhlan regiment in March, 1862. Soon, however, something went wrong there. On the birthday of King Wilhelm, on March 22, there was talk in the officers' mess – not surprisingly – about politics. Not a week had passed since the House of Representatives in Berlin had been dismissed and the cabinet of the »new era« had been replaced, thanks to the conservative minister Hohenlohe-Ingelfingen. The constitutional question of the budgetary specialization of the expenses of the army had become acute. The conversation about this was lively. In that circle everyone was glad that now truly conservative men would shape history. One of the lieutenants asked Dohrn, a noncommissioned officer, whether he, too, was conservative. He answered: »I admit that I am for the progressive party.« There was embarrassment, belated questioning, and reproaches. It is stated that the political conversation was neither intended nor started by him; nevertheless . . . the captain, who liked him, saw to it that things would not get worse – should the noncommissioned officer report to another regiment? This was discussed and rejected. The cause of his giving notice would become generally known – nothing would be gained. He wrote to his mother that she should not worry, that »of course« he would not be an officer of the reserve. He simply had to get over the remaining few weeks; he was assured that the attitude was not meant against his own person: »My best consolations are the Pickwickians.«

The final report, however, of September 24, 1862, was rather brutal: »With his meager preparation, his poor talent for executing the service, and his entirely unmilitary behavior, he has not acquired sufficient knowledge to be promoted to reserve officer. Whereas, during his time of service he was not guilty with regard to good behavior, it must nevertheless be mentioned that he openly embraced the progressive party.« He was angry about the outcome of the matter, for he had no reason to reproach himself with anything and, at the age of 21, he did not wish to feel like a political martyr. Of course, his military ambition was minor, and when in 1863 he survived a serious appendicitis and his doctor forbade him to ride horseback, he left military service. In the wars of 1866 and 1870 he returned to the service, but in a subordinate position.

In Jena the anatomist Carl Gegenbaur was growing in importance. In addition to him, in 1862 the 28-year-old Ernst Haeckel appeared and quickly gained the limelight, not only in academic circles. Both men were to leave their mark for a long time on the medical and philosophic faculties of the University of Thuringia. They attracted the attention of the young Dohrn and determined his development. He learned physics from Schäffer, who had also been the first teacher of Ernst Abbe. Gegenbaur taught Comparative Anatomy, and Dohrn took part in the exercises of making preparations. Haeckel, with whom he studied Zoology and Histology, confirmed »his most excellent diligence and success.« They became better acquainted when, during an almost fatal illness of Dohrn, the teacher became the person in charge of his medical care.

He reported to the family about the state of the sickness and, in the opinion of Dohrn, truly saved his life.

Haeckel's lectures on Darwinism had a decisive influence on Dohrn. Haeckel was to see later that Dohrn, as he matured, learned to look skeptically at his popularization of the English scientist. But at that moment Haeckel helped him to overcome his inner crisis (he questioned whether zoology was his true field) by introducing him to Darwin. Doubts about this had bothered Dohrn several times. He almost left zoology to become a book seller. As late as 1865, he wrote in a letter that it might have been better if he had buried himself in history and social science, had not his father wished to raise the name of Dohrn from dilletantism to science.

The academic kind of zoology had disappointed Dohrn. Now, through the theses on the origin of the species, he found a newer and wider perspective. Fresh and unprejudiced, he soon supported Haeckel by publications. Haeckel should not, in a lecture in Stettin before scientists, »abstract from Darwin,« he wrote in September, 1863. »I informed the press already,« he continued. The professor, only 7 years older, in return offered the more intimate form of address of »Du«* to the student. For Dohrn, who himself set limits to intimacy and held to them despite all sense of friendship, it was later a strange feeling to have shared this intimacy only with a man from whom in later years he separated himself completely. But the letters of his youth, to Jena, are full of a jocular intimacy and true participation in Haeckel's personal tragedy, the death of his wife.[11] The letters are full of extravagant plans: the student was afraid somebody might be ahead of him in presenting the theory of evolution and might reach a wide audience. There is talk of »hatred of the clergy and of the military« and the »mighty blow against the black bastards.« Even in Jena, science did not fill his life, he also associated with the journalists. »In the evenings,« he wrote to his mother on January 23, 1863, »I am now busy with writing political

* The German language distinguishes between »Sie« and »Du«, in older English »you« and »thou«.

articles for the local newspaper. Since they are rather sharp, they attract the attention of the readers . . . In Prussia, by the way, my articles would not be published. Don't tell father about the newspaper articles; he may not like it. For me they are very good exercises in writing and thinking.«

But his mother did not like these exercises either. Upon her strong objection he told her a week later that he took two articles back from the printshop and burned them; he did not want to make her unhappy. One can understand the content and tone of these articles from the defense he adds: »Concerning the content of these articles, you are mistaken. You can attack someone's honor only if he has it. Those who nowadays offer themselves to turn a blossoming country into a ruined one, to drive a cold-blooded people to revolution, have in my opinion no honor. That it will come to this, seems unfortunately only too certain . . .« It is the mood of the radical anti-Bismarck opposition which we hear in these words, which will continue for some time. With some irony, he tells her that he could not leave journalism alone and now had started a press campaign (incidentally a successful one) for starting nightly buses from Apolda and Weimar to Jena – he wanted to go to concerts. To this, his mother had no objection.

As before, his scientific work dealt with bedbugs. A prize competition of the medical faculty directed him to a new field in which he was to work much later, the crustaceans. It called for a microscopic investigation of the terminations of the motor nerves on the muscles of the crayfish: he was given the first prize. At the entomological section of the convention of naturalists in Stettin, the 22-year-old student gave a talk on hermaphroditism, his first public appearance. What he presented was the imperfect state of our knowledge, the insight »that our information about the circumstances of generation is extraordinarily limited and may remain so for a long time, due to the unfortunate physical orientation of presentday physiology.« Even Darwinian theory will find it difficult to explain these strange phenomena. Perhaps future research will succeed.

Dohrn intended to finish his studies in Berlin. He enrolled there in the fall of 1863 and stayed until spring of 1865. He registered in the medical school and passed his premedical exams in the spring of 1864. Only the state medical examination remained to be passed. His primary teachers were the physicist Dove and Rudolf Virchow, whose sympathy, both encouraging and critical, later became important for him. He did not relate well to the zoologist, Peters – when, in July 1865, he wanted to take his philosophical doctoral exam, he failed, and Peters adivsed him bluntly not to try it again for a whole year. This defeat hurt his lively self-esteem. It also put him into a difficult position within his family. Since May 14 he had been engaged to Therese Zitelmann. Haeckel, first to receive the news, learned of it the day after: »She has a pure, strong nature which, through me, will certainly develop in a direction that will follow my idealism step by step.«

His father agreed to the choice, although it seemed premature. Her family belonged to the larger circle of friends and also owned a country house in Hökendorf. But first the exam should be passed – then they could make their next plans. Anton Dohrn had written to Haeckel that he wanted to go to Cyprus after the exam and work there. Thereupon came Haeckel's suggestion to accompany him to Dalmatia. The father had agreed. All this was now disrupted. Carl August Dohrn started a correspondence with Haeckel: he should tell him his opinion on the entire matter. The description of his son is unflattering: »healthy, sly, and if possible not eager to work.« The young man should first reform himself before he tried to reform the world. Perhaps it was, unconsciously, also Haeckel's fault, because his self-confidence had led the son astray, without attaining any positive achievements. At the same time the son wrote to Jena – not knowing that he was being discussed there. He would like, nevertheless, to go on this trip and he promised to make good for his failure. Haeckel's answer to the father is quite satisfactory, although his criticism of the son matches the father's.

In the end, the plans for the Mediterranean trip were given up because of an outbreak of cholera: in August Anton accompanied

Excursion to Helgoland, August 1865. *Standing*, from left to right: Anton Dohrn, Richard Greef (Bonn), Ernst Haeckel (Jena); *sitting*, from left to right: Salverda (Delft), Pietro Marchi (Florence)

his teacher to Helgoland for 7 weeks – it was the entomologist's first encounter with marine fauna.

Dohrn did not feel himself to be quite without fault in the exam failure. Although he explained it by complaining about his bad memory, about the lack of result of painful memorization, he nevertheless wrote to Haeckel that his »polyhistoric tendencies« were also part of it. Under the professors who managed to attract him, he had worked with diligence, but the concentration on his main field had been endangered by this environment. In January, 1864, he had been in Paris for a few weeks under the guidance of his brother Heinrich.

Heinrich Dohrn [12] had followed his father more consciously in his effort to become a scientific collector. He had studied natural sciences in Bonn and Zürich, had been in London once in 1857 and again in 1861 – in that year he had also traveled to Russia. Almost always the visits were meant to meet the friends and correspondents of his father. He had helped him also in the preparations for the scientific meeting in Stettin. During these preparations he discovered that the city had nothing to show the visitors. At the age of 25, he founded the Society of Pomerania's Museums. In his long life he founded several other things, sometimes with luck and sometimes without skill. Despite his sickly constitution he dedicated work and conviction to politics and business – but in the unhampered work for a Pomeranian Museum he early found the center of his life. In his old age the city thanked him for his loyal service by making him an honorary citizen.

In the early 1860's, however, he was just a young man with a Ph. D., uncertain about a profession. The letters of the younger brother greeted the »Director of Pomerania's Provincial Museum«. This not yet existing institution was the cause for preparing a trip »to collect natural historic objects«, as it is called in the contract which Heinrich Dohrn made with the person who was to accompany him. The goal was West Africa and the Islands of Cape Verde. Before the start of this enterprise which lasted from 1864 to 1866, Anton visited his brother in Paris. Nothing of his impressions is left, but Heinrich reported to his father that Anton »was touched by the completely different conditions and the buildings . . . perhaps this trip makes him wish to see a little more of the world than Berlin.« Heinrich felt himself to be a globetrotter.

However, for Anton the Berlin of that period was world enough. That Berlin saw the rise of the constitutional conflict between Prime Minister Bismarck and the majority of the lower House. Ferdinand Lassalle* established the headquarters of his socialist propaganda there. Berlin also saw the army leave for the war with Denmark** and return a few weeks later with the fresh laurels of victory. In the mind of the young Dohrn the sensations were confused – he shared this confusion about the exciting period with many young men of his generation. Obviously, Lassalle's motivations and their aim moved him strongly. He wrote his father, who had »strongly criticized« Lassalle, an exact presentation of the »love and duel story which I heard from Bucher«. The very detailed story (which corresponds quite well with later historical examination) would not have been told if he had not wanted to correct his father's bad opinion. Dohrn did not get to know Lassalle personally; but even his particular witness indicated the circle to which he belonged. Bucher was at that moment on his way from Karl Marx to Bismarck.

The center of Anton Dohrn's stay in Berlin was the house of Adolf Stahr and Fanny Lewald;[13] both meant a lot to him. The close and intimate relationship, lasting over years, was in some way amazing – Dohrn himself found it amazing in retrospect. For the presentation of his life, particularly his years of inner crisis, it became incomparably important. His inborn need for self-expression found there the place where he met friendly understanding and good will – they patronized him and advised him in many difficult situations. They were themselves somewhat flattered that such a lively and flexible young man looked to them for – and probably

* Ferdinand Lassalle (1825 – 1864). Organizer of the German Labor Movement; influenced mainly by Hegel and Karl Marx. His goal was »Freedom of Mankind«. Fought in the 1848 revolution on the democratic side and was opposed to the bourgeoisie.

** 1866: this was more or less an internal German war during which Prussia freed itself from the Austrian Imperium in the battle of Königgrätz. Prussia also annexed the Kingdom of Hannover which was being changed to a Prussian Province. (The Kings of Hannover had also been Kings of England since 1714.)

found – a place of refuge for his restlessness. He might think of himself as mature and superior compared to his contemporaries. Here, he looked up gratefully and adoringly; he enjoyed the atmosphere of educated discussions, of pessimistic opposition, of those who saw in the present period only a fatal and doubtful interplay of political and intellectual history. He was too young to see the pretentious and vulnerable vanity in these two people. His father saw the relationship differently; he complained to Haeckel in February, 1866, about »the miserable influence, since their artificial intellectual wit has surrounded him with a fog of false apprehensions«.

In the Berlin of that time, Stahr had a recognized position as a publicist. He had published works in history and literature, and could be considered knowledgeable in the problems of Italy. Since he had written a successful liberal biography of Lessing,* he felt himself to be the advocate of his tradition. He was in contact with very many people and knew that his praise was desirable and his blame was feared. He had a kind heart, a pleasant, educated personality, but he made the half-hidden claim to have the last word in matters of art.

In 1855 he had married Fanny Lewald, who had won a wide audience with short stories and novels and in the early 1860's had just published her biography in several volumes. Her poetic gift was minor. She owed her success to an elegant and self-assured presentation, using people and human destinies as carriers of theses and tendencies – a repetition of the spirit with which the »young Germany« had tried to take the place of the Romantics. Her ambition was – and she was partly able to achieve it – to create a literary-poetical »Salon« similar to the one Rachel Varnhagen**

* Adolf Stahr: *Lessing, sein Leben und seine Werke*, Berlin 1859. Translated into English, Boston, 1866.

** Rahel Varnhagen, née Levin, like some other well-known women had a salon in Berlin, where the intelligentsia gathered.

had created for the preceding Romantic generation. Gottfried Keller,* too, had entered this circle and had disgustedly turned away from it. His grumbling words of the »four-legged, two-sexed ink-animal« accompanied the Stahr-Lewald couple into their somewhat paper-thin immortality. Nevertheless, Fanny Lewald was not unimportant as a contemporary phenomenon, although her poetic fame had already faded during her lifetime and nothing she wrote outlasted her. But her autobiography was a source of knowledge for the emancipation of the Jewish middle class, and to a lesser degree also for the social independence of women.

When Dohrn, at the age of 23, entered this circle he was fascinated: »the loyal Anton Dohrn comes every day; also Dr. Johann Jacobi, who is still here, honors us with a daily visit.« Stahr wrote this on July 7, 1864. The tone of Anton's letters to the couple is warm and intimate. He was eager to introduce his friends to the Stahrs and reported to his father about them – he only mentioned ironically Stahr's complete lack of talent in card playing. Later, he recommended to some English acquaintances who knew German that they should not miss the novels of Mrs. Lewald – as if they were important products of the German mind; he sent them to these friends overseas. He felt a warm affection with a trace of youthful enthusiasm for them, which he did not normally have; he once called the couple »his true parents and most loyal friends« and characterized himself as their »intellectual son.«

Later, in the mid-1870's, when he himself had become more sure of himself and the world, more experienced with people and had developed a sharper eye for problems, he made an almost cruel and hard confession in a very characteristic letter to his friend Grant: »Fanny Stahr is back here again. You know that Stahr died. If I ever felt that I dislike her, it is now the case. Her heart and feelings are made of skittle from which bowling balls are made. It is almost

* Gottfried Keller, Swiss author of the nineteenth century, who wrote cycles of short stories and some novels, most of them concerned with Swiss folk-life.

more bearable to see her decidedly cold and hard than occupied with feelings whose depth and power she never understood nor owned. There is no doubt that the death of Stahr changed her life and made it poorer, which is no wonder with his rich emotional life, the sometimes unbearable, but often charming moods, his incomparably more delicate esthetic feelings which characterized him, in contrast to her. I am also personally sorry for her. I have the feeling as if this woman had suffered for the first time a real pain, as if she had never known the inner death, the stir of yearning, the helpless struggle to find a different reality, a sense of an envisioned, but unachievable perfection – in other words of a real soul capable of great feelings. What an incredible difference between her and Rahel, as great as the difference of the two periods of the German mind, in which both live and lived. I can barely say how my whole being now turns against her. In Stahr I felt almost every movement of my own being. I disagreed with much, but I understood his stupidities and realized the necessary limitations of his nature. But how I could ever have loved her? I probably never did. Stahr I definitely loved. But Fanny commanded me because I was weak – now I feel my independence from her and can understand that we are not real friends. But I do not for a moment doubt her importance. You know all of this better than I do. Let us forget about her.«

3 JENA, 1866

In a general reflection written in the 1870's, Dohrn contemplated: certainly, the average educated person, when asked about the significance of the year 1859 in world history, will refer to the Lombardic War, the end of the Pontifical State, the mighty step toward the unification of Italy. Nobody should fail to appreciate the importance of these events. But the year draws its actual historical importance from the publication of *The Origin of Species* by Charles Darwin. This work meant to Dohrn a revolution not only for the methodology of natural science, but he saw that in the Darwinian doctrine, if it were exhaustively explored, the continuity between Nature and human history was secured. And he expected from it also the most profound influence on the approach and the practical consequences of social economics. One cannot value the impression made by the English master highly enough, particularly during his decisive years. More than once Dohrn was inclined to despair of his science and of his own potential in it. How did he look at the situation? The first issue of the *Entomologische Zeitung* of 1866 contains in a review a few sentences by Dohrn which – one is not sure – might have hurt the parental editor: »While we encounter a truly immense extension of the purely classificatory-descriptive works in our science, the anatomical-descriptive, the embryological, and the physiological works can be considered as scanty. The interest in collecting directs the path of science. We shall always be forced to assume a temporary stability of the species. The system and the nomenclature must serve us as a necessary exchange of communication, to make our observations and discoveries communicable and comprehensible. But we can no longer entertain the idea of considering our system as the expression of a plan of creation, nor the identification of a new species as the foundation of an idea of creation, since Darwin has deprived these conceptions of their last remnant of plausibility.« Already in July 1865 he had written against the pure systematists: »The forces which hitherto have been used for the separation of species will have to be applied to the opposite, that is, the unification, the derivation of the one [species] from the other.«

Jena was the place where the inroads in the traditional field of zoology against which these statements were directed had the most radical effect. It was obvious that Dohrn should turn there, if he intended to pursue his earlier project of a University career. In the early summer of 1865 he asked Haeckel to accept him once more for 2 years as his student; medicine was no longer a consideration. Haeckel agreed to this proposal, although not too enthusiastically. Perhaps he noticed in the enthusiasm of the letters the differences of their personalities. In November of 1865 Dohrn had obtained his Ph. D. from Grube in Breslau without any difficulty. The topic was still derived from entomology: *On the Anatomy of the Hemiptera.*

At the beginning of the new year he went to Jena. Very soon, the relation between Haeckel and Dohrn became problematic – the change from close intimacy to hurt feelings is reflected in Dohrn's letters to Adolf Stahr and to his father. Both had strong reservations against Haeckel. Presumably, the younger scholar, who liked to throw the gauntlet to all authorities, had not appropriately appreciated the old Stahr on his Berlin visit, so that the latter had downright warned Dohrn against Haeckel. Dohrn defended Haeckel's human sincerity most animatedly; he even found some merit in Haeckel's handling him so harshly. In a letter to Stahr (April 19, 1866) he said: »Here (that is, in a rural petty bourgeoisie) Haeckel finds the ideal of his world. He calls every polite, considerate behavior a »social lie,« and he attends tea parties in muddy, filthy boots. But I have no accusatory sentiment against him. He has given me the utmost in friendship, within the limits of his personality, and I have the duty to reciprocate ... And, as you notice quite correctly, I perceive myself as the superior. I shall find ways and means not to give this splendid but limited man the right to accuse me of ingratitude of the worst kind ...«

It was unwelcome that Gegenbaur[14] objected to his habilitation,* but that was not very important to Dohrn. Through his ungentlemanlike polemics against an old zoologist, Haeckel had hurt his feelings. He had importuned Dohrn »not only with merciless but with downright murderous words.« On the eve of his appointment as Privatdozent, he made the self-critical statement, »I really have no calling as a zoologist.« And he explained to Adolf Stahr how he had been convinced earlier in the »ways of the Dohrns«, that zoology was really *the* science of sciences ... »But gradually I became amazed by the boundless lack of ›Bildung‹** of

* Admission to untenured faculty status.

** Culture, Literacy.

the zoologists, and I have become quite sure that culture does not belong to this guild, that it only does damage in that it cannot combine its demands with the requirements of zoology.«

He had been stamped as a zoologist already in the cradle, no matter whether he was suited for it or not. Much later, however, when he was hit by »the very deeply moving excitement of Darwin«, he dedicated himself to zoology with all his strength. »Even if the practice may be tedious and dull, Darwin and his theory give it a relief that could not be more grandiose.«

At that time, Gegenbaur and Haeckel had made it clear to him that he was lacking in the great moral »strength« for zoology, the »calm perseverance«, for which he was not concentrated or industrious enough. He himself, aware of the necessity of scrupulous observation, of careful, constantly repeated experiments, became worried about whether such resignation would be in accord with the calling of his life, of which he had a fairly high opinion. At any rate, he would not capitulate now. If it did not work out with zoology, then perhaps with anthropology. For a while it seemed as though that might save the situation. But a closer consideration of this possibility showed that even though he had already gained ground in this area, which seemed to offer more fruitful nourishment for his imagination, the preparation would require more years of study, with which he could not and would not burden the old man in Hökendorf, who had gradually become impatient with his ambition for his son, and who, in February of 1866, had suffered a disappointment when Anton had broken off his engagement. Dohrn had to justify himself thoroughly about that before Haeckel. But he realized that he would make the attractive girl unhappy.

If he had received more admiration in the Berlin circle than he possibly deserved, he sensed in Jena that he was surrounded by a skepticism which could be overcome only by achievement. The actual entomological period was concluded with his doctoral thesis. It lingered on in the almost daily intercourse with one of his father's friends, the deacon Schläger in Jena, whose humanity and collector's zeal he valued more highly than his sermons, which he attended several times out of politeness. Since his prize-winning work on the river crayfish, and since his visit in Helgoland, Dohrn had turned to the crustaceans. He worked on the water woodlouse, and in his reports to his father he outlined the plan to write a genealogy of the crustaceans which had been neglected by science for several decades. Above all, this heralded a turn to embryology and with it a concern with the problem of when, where, and how to find the right material. He also wrote to his father about his difficulties with the professors in Jena; and he strongly recommended the book on Winckelmann by Justi, which had just been published. This shows how broad the young man's scope of interests had remained while he occupied himself with woodlice.

To be sure, there were other reasons for finding it difficult to work quietly with the microscope in the year 1866. While the constitutional conflict in Prussia between the minister and the chamber of deputies continued without relaxation even after the Danish campaign, Bismarck had started to set the German problem in motion. The young scientist was deeply excited and gripped by the events and even more so by the questions that arose in their shadow. The political views with which he had lived through the beginning of the 1860's contained very many elements which were valid for the times and for his generation, although in a very personal mixture. The family tradition of aristocratic rejection of the philistines and demagogues belonged to the middle-class liberalism which was closely connected in these decades with the interests of harbor cities. The house of Dohrn had, from his father's side, the fresh air of a free internationalism. The personal friendships reached out into many nations; a kind of cosmopolitan openness prevailed. If Anton set himself off ostensibly from the tradition, it is not in this particular respect. His utterances of the early period are characterized by social radicalism, and an antibourgeois attitude. Now, Carl August Dohrn was certainly not a »bourgeois.« If one accepts the meaning of this notion, he was much too indifferent to all matters of acquisition; he was a late romantic. But his egocentric individualism closed his ears to the social problems which intruded at that time; with patriarchal calmness he carefully settled the immediate obligations which he encountered, and everything else was of no concern to him.

But his son had been gripped by just that, and with a certain vehemence. He had been vividly excited by the adventures of Lassalle, and had tried to gain a clear idea of social economics. This went on for several years, and he asked himself repeatedly whether he should not find his life's goal in political hustle and bustle. To be sure, he was not made for the rough and coarse side of political fighting; of this he was perhaps aware. But the political elements of his talents were to develop and to unfold later in a most unique and successful way, that is, in argumentation with individuals, in the knowledgeable handling of people, and not in compact encounters with masses and the fate of the masses. This moved his sensitive feeling for justice. But the cultivated, differentiating need for distance and for a purely intellectual dialog separated him from the demagogical mission, from the calling of an active friend of the people, from a public fighter.

Now, in the spring of 1866, he did not find it easy to embed his oppositional feelings in the historical situation and concerns. The letters to Stahr and Lewald reflect the peculiar course of his sentiments. On May 22, when the news of Prussian-Italian negotiations transpired and the diplomatic struggle in the federal Parliament in Frankfurt began, he inquired whether one did not »talk in Berlin of more effective measures against the government«. »What does the old Johann Jacoby say and think?« A few days later he considered war against Austria, »against this caricature of a State«, as inevitable. But the revolution seemed to him as »by far the lesser evil, which might lead to better circumstances«. He grew accustomed to the idea that »civilization experiences a bloody interval«, and, bent over his manuscripts, he armed himself with »coldbloodedness for the less quiet affairs that are to come.« Then he examined what one could wish for the German people: whatever went wrong in Prussia was »not as bad as a victory for Austria, with its flood of reaction and medieval attitudes. I honestly confess to you that

I would joyfully let go of everything to fend off this fate from the German geople with all my strength.« Then followed vehement outbursts against the South Germans and particularly against Bavaria. »The Ultramontan* doings and drunken rudeness are worse than any trait in the Prussian character. I hope and wish for the Prussian victory with all my heart.«

These are, in part, monologs, in which he is thinking out his arguments. If one finds at the beginning a certain sympathy for the smaller states, he also came to terms with some well-planned annexations »which deprive the world of some bloodsuckers.« He had a sense of being an »unbridled and reckless republican.« In fact, he was not »that much disheartened that a cesarism is in store for us which would kill freedom«, because, in opposition to its reality, the republican ideal would grow in many hearts. His friends were dismayed that he prepared for participation in the war, by horseback riding and shooting. He explained to them his inner necessity: »On whom do I wish to have an effect? On young men, who above all demand of a man personal courage.« And furthermore: »Unless *we* elevate the war it remains only a dynastic slaughter; through us, it can acquire an idea.« When he had again donned a uniform, he found himself »in a more bold and warlike mood«. He found that »there is much more strength in it than in the eternal black and gray«. »I have always known that compulsory military service is democratic in the best sense and now I won't stand back but shall fulfill my word.«

In these weeks he worked up a vivid resentment against the »Habsburg robber dynasty.« He still hoped, when the first great decisions had come to a head in Bohemia, to witness the great moment »when cultural history takes a gigantic step . . . When I ride into Vienna as a green hussar.** I shall take an oath not to become a philistine, and I shall never forget that.« He was drafted to the 11th Hussar regiment in Düsseldorf. But he did not go to the

---

* Reactionary catholicism.

 ** »Green« often means »inexperienced.«

front any more. A few weeks later, he sat very peacefully at the estate of his brother Wilhelm in Bahrendorf. In his letters, the warlike and political strains were completely blown away. The moving theme is now F. A. Lange.

4 FRIEDRICH ALBERT LANGE[15]

In a letter to Stahr of April 27, 1866 Dohrn mentioned for the first time Lange's *History of Materialism* * which badly shook up his own up to then »rather harmless« materialism. The book moved him deeply; it also led him to a systematic study of Kant: »Not long ago I bought the entire *Works* by Kant, and I am enthusiastic about the marvelous calm and clarity of this wonderful man. The optimistic, inoffensive tone of his opinions and convictions is very beneficial.« These sentences were written among some excited letters a few days before the battle of Königgrätz. One can feel how the constraint of abstract thinking became shelter and protection for this young lively mind and his passionate soul. But this man Lange was certainly not »inoffensive«. Dohrn did not know more about him than that he was »a penalized teacher of the gymnasium at Duisburg who had refused to leave the National Society on command«. Such daring seriousness of life was enough to impress Anton Dohrn. But the book fascinated him; he asked that everybody should read it. Nobody knew the book, which had come out late in 1865. He wanted to testify in public for the book and the man. On August 30, 1866 he wrote to Lange: »It is not only selfishness which makes me wish to contact you by letter and possibly to make your personal acquaintance. I am too profoundly indebted to you to simply accept peacefully the extraordinary quality I enjoyed in

* Friedrich Albert Lange (1828–1875). *Geschichte des Materialismus und Kritik seiner Bedeutung in der Gegenwart* (Iserlohn, 1866). [History of materialism and its meaning for the present time.] Considers materialism as a partial function of scientific method. A follower of Kant.

you, without telling you how much I esteem you since I read your *History of Materialism*. You succeeded completely in conveying the purpose of your book as you suggest on page 241. In fact, your book carried me away and I must confess that I have never yet met such a complete blending of noble humanity, acute and honest thinking, widespread knowledge, and harmonious beauty of expression as appears in this book. I was a materialist, glowingly convinced of the supposed truth of my convictions, which I felt not to be open to attack. I am defeated by your book. But instead of polluting my environment – like a corpse on the philosophical and scientific battlefield – and polluting its inhabitants with miasma and decay, instead of casting anathema on all speculations (as most of today's »exact« scientists are still doing), your book has seized me and forced me to wipe away the old bad growth and let it penetrate and renew me. It is in my nature that I am easily inflamed and, out of a profound need, open to enthusiasm.«

F. A. Lange was not used to receiving such letters, particularly in that period: he was in the midst of some annoying law suits because he was supposed to have offended the Royal Superior Court. He was ready to build a new existence for himself outside of Prussia, in Switzerland, as a publicist or scientist, whichever came his way. He seized the hand which a stranger stretched out to him, both touched and proud. He opened up to him, talking about his troubles. A very intense correspondence ensued, which was carried on for several years. In April, 1868, they met personally in Winterthur. The contact with this man caused a revolution in Dohrn's view of the world, as he often gratefully acknowledged. It confirmed his socio-political instinct and furthered the economic insights which he was trying to clarify at that time.

Lange, 12 years older than Dohrn, belonged to the strange sideline figures of the intellectual and political history of Germany in whom great talents were latent but did not come to fruition, ending in the shadow of tragedy. His second letter sent to Dohrn in October, 1866, gives a part of his preceding biography: »You conjecture rightly that my many-sidedness is in part the consequence of my fate. Many times I have made virtue out of necessity. For I was a professor, teacher, editor, secretary of the chamber of commerce, book-seller, insurance agent, etc., aside from numerous side jobs.« This enumeration could have been increased by some variants a few years later. Lange had entered Swiss politics, unions, and popular education. Finally, Zürich gave him a teaching position, and in 1872 the Prussian minister of education Falk called him back to Prussia. Although he was fatally ill, Lange took over the professorship of philosophy in Marburg; he died not long afterward in 1875 at the age of 47 and left behind him a renewed interest in Kant's philosophy. More than once, in observing his fate and his way of thinking, one is reminded of Friedrich List.* It is the life of an outsider who destroys himself between his striving for practical effect and the passion for nonprofessional thinking, with a soft soul, but harsh in human relationships, a man threatened by bitterness who dared to stand up proudly for his inner freedom.

One cannot follow the traces of Lange's influence in Dohrn's later works. His tasks, when they were of an organizing nature, were governed by a changing and peculiar situation, and the specifically scientific work followed certain methodological presuppositions, the correctness and need of hypotheses and deduction. In later years Dohrn's literary production paid no further attention to questions of the theory of knowledge of scientific research. But in that period of youthful crisis, which was not only a personal one, it was for him of decisive importance to have gone through this intellectual self-examination as a follower of Lange.

It gave him a sense of security (which soon became of great importance) as he opposed some of the scientists, Haeckel among others, out of a feeling not so much of superiority, but of intellectual freedom and broadness; and it gave him the strength to see the limits and remain free from the hubris of the self-assured guessing

---

* Friedrich List (1789–1846). Theory of economics and politics. Worked for industrial capitalism.

game of science concerning the world, at a time of scientific enthusiasm – which later was also supported by his own work.

F. A. Lange did not leave a »system«; he actually objected to those who expected one from him. His historical importance lies in the fact that a philosophic mind tried to close the gap between philosophy and the exact sciences. Speculative idealism had erected a grand monument of conceptual abstractions with which it hoped to overcome reality. But it no longer reached the young generation, whose efforts were directed toward the knowledge of nature, not to anything else; for it, unproblematic positivism was the easy way toward materialism. The number of scientists who tried to find the presupposition of a critique of reason for their work was small. Here Lange started, after having thoroughly analyzed the questions and answers of science. He chose a historical way; and this trip through the history of mankind gave the work its peculiar charm. With an open-minded sense of adaptation, it sorts and evaluates, recognizing the importance of periods and opinions, without becoming vehement about the global judgments of the past.

When Lange approached the present, one feels an element of tension and, once in a while, a happy desire to fight, which reached Eduard von Hartmann* as well as Moleschott and Büchner. Here he did not want to teach, but to influence.

Anton Dohrn acknowledged this influence not only in the letter, but in the great essay which was published by the *Nationalzeitung* in Berlin (October 31, 1866). It is like a fanfare. Lange did not hide from him the fact »that in the beginning I found it slightly unpleasant that you had praised me so extravagantly«. He would have preferred the essay to have entered more deeply into the

* Karl Robert Eduard von Hartmann (1842–1906). *Philosophie des Unbewußten* [*Philosophy of the unconscious*] (eight editions). *Kategorienlehre* [*Doctrine of categories*], 1896, his most important work. *Weltanschauung der modernen Physik.* [*Philosophy of modern physics*], 1902. *Das Problem des Lebens* [*The problem of life*].

subject matter. But then he realized that »the fresh way in which your feelings are expressed is a certain guarantee for the effectiveness of my presentation«. Lange's publisher had told him that the sale of the book in North Germany had »greatly increased« after Dohrn's essay. Anton Dohrn said in his answer that he felt »too insecure« for a matter-of-fact criticism. With a good conscience he could write only as he had done: »It was myself that I laid open.« In the essay he had written: »That he is by profession a scientist and has enough anatomic and physiological knowledge not to be shaken off his materialistic opinion (which up to now he had considered infallible) by any kind of metaphysical talk. However, he feels completely shaken now. When he was attacked on the same ground on which he stood with the same weapons he had learned to use, when a philospher who was also a scientist appeared, at that moment he was defeated. Now he looked without pain on the error which was now overcome. For he considers an honestly defended error a necessary preparation for the knowledge of truth – in whatever form it may consist.«

Several times the correspondence touches closely on philosophical problems. The eager questioner who, with his friends Snell and Abbe, could not quite come to an understanding of the interpretation of Kant's »thing in itself,«* received an answer many pages long. Lange explained with great acumen that he thought of the »thing in itself« as an analogy to sense impressions (as long as he was not going over to a poetic metaphysics). Teacher and pupil tried, by a cautious conceptualization, to exclude possible misunderstandings, and to understand each other, in their view of nature, about the nature and meaning of causality, and the correctness and limits of teleology. But the biographical explanation of these doc-

* Das Ding an sich (The thing in itself). According to Kant a thing which is not subject of our sense impressions. It rather lies at the bottom of all experience, but cannot be seen or heard or felt. Much discussed later, mainly by Hartmann as the »idea« underlying experience. Generally rejected since the 19th century.

uments goes further. It is Lange's many-sidedness which brought him closer to this enthusiastic student. Dohrn was himself uncertain about his ultimate goals. But the master was not to disappoint his questions. There are Lange's writings about the workers' problems, about John Stuart Mill and Carey. Dohrn threw himself with a certain passion into socio-economic studies, having just declared that he would try again to understand Schelling's * philosophy. But Lange's violent opposition to the economic, vulgar liberalism of the »Economic Congress« touched him. He wanted material to get to know more closely »the economy of egotism« which had to be fought. The fact that Lange led him to Mill in one of his writings, was for him at that moment a »great gain.« He accepted the author of the treatise *On Freedom* among his »saints.« It is out of youthful enthusiasm, that he wrote to Lange (October 18, 1866): »You are musical, aren't you? Could you feel with me when I compare this work with Beethoven's great quartets? I mean Opus 127, 130, 131, 132 in C sharp minor, B major, A minor, and E flat major. For me, they are revelations of the eternal realm which I, in those moments, feel to be the right of those suppressed here. The Zeus of Phidias, who dominates my room, the great quartets, and the first movement of the Ninth Symphony are for me the most overwhelming creations of art: the purifying effect they have on me, I have now experienced in Mill's book. Therefore, I take him along with Goethe, Kant, Beethoven, Schumann, Mozart, and Phidias. Don't laugh at me on account of this gallery – they are my gods and I am, in their service, very orthodox and intolerant.«

A year later, he said that if he were rich he would give the book as a present »to each member of the House, so that this miserable chitchat of facts and practice, in contrast to theory, would finally stop.« He contacted Mill and asked him for permission to translate

---

* Friedrich Wilhelm Schelling (1775–1854). Professor of Philosophy at various universities, mainly Munich and Berlin. Considers the world as continuous history of self-consciousness. *Philosophy of Nature* (1802–1809). Married to the former wife of August Wilhelm Schlegel.

into German Mill's *Address to the Andrews University*. Lange doubted that the presentday Germans were able to accept it. Nothing came of this plan. It is not known whether Dohrn, when he warmed up to the English thinker, knew of the great confrontation, in which, a few years earlier, Heinrich von Treitschke* had attacked the pamphlet on freedom.

Years later, when he sent Lange his programmatic essay about his growing life's work from Naples to Marburg, Dohrn wrote (September 26, 1872): »I hope you do not expect that I send you again, as formerly, some indigestible philosophy, as at the time when the History of Materialism had completely shaken me. I have known for a long time that I am not fit to be a philosopher; but of course it took me a long time to put myself back on my own two feet in the face of that whole way of thinking. Nevertheless, there remains in me a large sediment from that period which had a decisive influence on my later life. For one thing, it led me to a special, rather large scientific task, on which I am working independently, but which will not be published for another 2 to 3 years. For another thing, your chapter, *Ethical Materialism and Religion*, more specifically your words, ›Wealth is a task!‹, exerted on me the strongest influence and became for me the *motto* of my existence.«

5 THE ROAD TO BECOMING A PROFESSOR

In a letter to his father dated November 23, 1866, Dohrn wrote: »Sometimes I get dizzy when I consider in what an age of revolution we are living! Geology, morphology, physiology, anthropology, all completely reformed, history in the throes of birth to assume a new image, economy and statistics just born, meteorology created; and on top of it all the twilight of a universal-scientific educational ideal, which is revived in many young minds, but also in

* Heinrich von Treitschke (1834–1896). Historian, very reactionary. Defender of Bismarck. Prepared the imperialistic ideology.

many old leading minds – if this isn't a fermenting age, there never has been one. But the opponents are strangely distributed at this moment. The radicals together with the reactionaries – both with the knowledge of profound hostility which will break out as soon as both have defeated the enemy – in my opinion their common enemy is liberalism in all its shapes from the Old Liberals to the Party of Progress. But your view of the situation of the world is different. The sad thing is that I fear that these different views will not find a peaceful union and therefore will make a reform impossible. But then a terrible catastrophe will result – that is inevitable.«

Such writings were not particularly welcome in Hökendorf. The old man disliked the violence with which Anton fought for Lange and made some harsh attacks against well-respected names (such as Schleiden). The son had to accept the rebuke. After all, he knew that his father preferred that his son write about the progress of the woodlouse story or send him entomological contributions – which happened from time to time. But the study on woodlice did not really progress. A study on the development of phryganeids worried him, because the Russian, Ilja Mečnikov, began publications on insect embryology and he was afraid that the independence of his own research might not be recognized. How would these things influence his »habilitation« as a professor?

The relationship with Haeckel, with whom there had been misunderstandings in the previous Spring, improved again in the winter of 1866–67 – there was an exchange of ideas and a »reconciliation.« Haeckel wished that Dohrn would discuss his recently published *General Morphology*. Dohrn agreed, admiring the enormous power of concentration of his friend, and was impressed by the unified penetration of the great subject, and yet repulsed by the self-assurance of the philosophical position which went far beyond zoology. In the letters to Lange, to his father, and the Stahrs, Haeckel was constantly discussed. He considered Haeckel to be at the top of German zoology, particularly in comparison with other systematic attempts; yet he foresaw his defeat in matters of philosophy. He wrote to his father: »The philosophical chapters are rather bad and written with the pretentious assumption that he wanted to show the scientists who understood so little about philosophy, what philosophical education could achieve. In this respect he has failed completely. I am afraid that if a trained person wanted to kill those chapters, d . . . d little will be left of Haeckel's *Monism* (November 10, 1866).«

Wilhelm Wundt carried out that attack as anticipated. For a while, Dohrn considered the idea of writing a long essay about this conflict, but the work was not printed, nor was the manuscript preserved. Haeckel knew it. Dohrn wrote to Stahr in July 1867 that he »was sorry not to see my review of his book printed, even though it contained such severe criticism, because I still understand better than anybody else what new and good things the book contained.«

In an essay in the *Entomologische Zeitung* (1867), Dohrn wrote that through Darwin the entire morphology would undergo »a complete revolutionary change and a giant step would have to be taken in that science. Before anybody could expect it, this giant step – in large dimensions and great many-sidedness – was taken by Haeckel.« He had told his father that he would like to avoid philosophy because it was evil – but that he would try »to cover his (Haeckel's) back.« In the correspondence with Haeckel Dohrn very freely discussed the flaws he saw.

Jena offered certain difficulties to someone trying to gather research material – Dohrn had complained about this several times. In early April of 1867 he went to Hamburg to work there in the Zoological Garden. This stay became important for his later work because he met the Englishman W. Lloyd, who had just built an aquarium there. He was to be of important assistance to Dohrn's own efforts. The president of the Zoological Garden, Dr. Alfred Meyer, a merchant who did scientific work, became close to him and took him along to his summer house in Kiel Bay, whose fauna he had already investigated. Anton Dohrn spent some happy months in this educated and broad-minded environment, which he also liked on a personal level. The study of the development of the crayfish progressed, and his letters are less restrained. The much admired singer Stockhausen and the composer Bruch were people who fascinated him. Will Hamburg keep him? Meyer asked whether Haeckel or perhaps Dohrn himself might wish to take over the position as director of the Garden. It became vacant because Alfred Brehm, the author of *Animal Life*, had decided to go to Berlin. The conditions were good – aside from popular lectures, an obligation to work scientifically was expected. A thorough correspondence with Haeckel tried to make the position desirable. Haeckel did not feel comfortable in Jena, but he refused. For Dohrn, the temptation was great; his friends were of different opinions. Lange, who had a great opinion of the educational value of such gardens, tended toward a positive decision. Stahr dissuaded him; he disliked the idea and was somewhat worried that it would make Dohrn evade his real task. Siebold and Leuckart were asked for their opinions, which were favorable. So was the judgment of Weismann in Freiburg, to whom Dohrn attributed the only and complete competence. The matter had one drawback: Meyer was on bad terms with the board. Dohrn's candidacy might easily become a proof of power for the president – Dohrn wanted to avoid this situation and rejected the offer. This interlude, however, had an unexpected, but not unpleasant result: Gegenbaur and Haeckel may have felt that their restraint in the early summer of 1866 had some influence on Dohrn's willingness to forget his academic ambitions. Now, they complimented him: the procedure of habilitation may be taken simply as a matter of form, as a topic of conversation. It should take place during the winter semester. Already in Spring, they had talked about and prepared a trip to England. The entomological friends of his father who had also received several visits from his brother Heinrich, took the initiative, particularly the ever-helpful Stainton. But this trip to England in the summer of 1867 was meant to investigate marine fauna; it became most important for Dohrn, although not decisive. It brought him in contact with a number of the leading English scientists and became the foundation of some personal friendships which meant not only a personal enrichment, but an important support for his later work. He was familiar with the intellectual life of the period in England, beyond science. Darwin and John Stuart Mill had meant much to him. He was advised by G. H. Lewes, the biographer of Goethe and a popular philosopher, and he loved the novels of George Eliot. With both he later had some personal contact. The young German was well received, he came to know English life from its most pleasant side, namely hospitality. In the British Museum he was told that he could reach his goal most easily if he turned to Mr. Robertson of Cumbrai Isle. Therefore, Dohrn went immediately to Scotland, in order not to lose any time. The professors in Edinburgh sent him on to Glasgow and he ended up at the house of this odd man, who offered him now and again in 1868 the use of his home and his equipment for months. Dohrn often talked about him with gratitude.

David Robertson was a self-taught man to an even higher degree than Meyer of Hamburg, whom he had just left. As an orphan, he had tended sheep on the beach and had at that time started to observe marine animals and to collect them during ebb-tide; he later progressed to tending horses, but his mind remained directed toward natural science, which led him to a position as assistant in the house of a doctor. But he realized that the way to a university could not be found in this manner. From the rest of his savings, and with some credit, he bought a china shop in Glasgow harbor, worked hard at it and eventually, after 12 years, the shop brought enough money for him to turn to his old love: in Great Cumbrai Isle he built a little summer house in order to watch and collect marine animals in peace. Robertson's wife became infected with this quiet passion. The atmosphere of the house agreed with Dohrn, he admired the open, intelligent mind of the man and his past. He wrote to Fanny Lewald that he hoped she would write a novel on such a character. The lovely landscape, craggy according to season, offered all desirable material. Robertson knew the places well where material could be found, they discovered much during the day or the night. Dohrn's letters breathe the pleasure, even of the physical strain. At the same time, he realized that he stood before a large scientific task which he approached with great assurance. His earlier work on the woodlice, the new investigations on larvae and embryos, seemed to him to lead to a unified genealogy, to which the newly started and recently published foreign studies on the development of embryos of insects and all articulates also seemed to point.

The German zoologist Fritz Müller, who lived in Brazil, was the first, in his book *For Darwin*, published in 1864, to try to apply the theory of evolution to the diverse forms of crustaceans. In the end, he suggested cautiously that the insects might have a common ancestor (with the crustaceans) in a typical crustacean larva, the *Zoea,* »which elevated itself to life on land.« Haeckel had adopted this notion in his *General Morphology*. Dohrn thought that in the early development of the crustaceans which do not have the *Zoea* larva, and even in that of scorpions, spiders, mites, and insects, he had found rudimentary indications of the structures of the *Zoea*, which seemed to reveal itself as the common ancestor of all articulates.

In a letter to Adolf Stahr of September 22, 1867, shortly before his return to Germany, he showed the nonexpert the importance of this: »If you want to have an idea of my kind of work, imagine that comparative linguistics had not yet discovered the key to the Indo-Germanic languages in Sanskrit; and that although the Scandinavian, Romance, Celtic, and other languages were fairly well known with regard to their inner structures, we still did not have any knowledge of their common ancestors, and could state, but not explain, their many similarities. Now, one would find in Sanskrit the root of the entire Indo-Germanic family of languages and the explanation of their origins. This I have done for crayfish, spiders, centipedes, and insects. The largest section and the richest in forms of the animal kingdom is thus explained, i.e., subjected to Darwin's principles, according to which the transformation of the forms of life proceeds.« However, the letter also contains the remark: »Most scientists won't believe that I obtained these great results. They consider the task to be so difficult that decades would have to pass before one could hope for a solution.« But Huxley had said to him: »Thunder and lightning! – this is the most important thing I have met with in a long time. You will start a revolution!« He had written to his father at that time: »Now I have the means to make a total revolution throughout the entire methodology and the entire field of zoology and to change Darwin's theory to Darwinian practice.«

The suggestive eagerness of the young German made enough of an impression to make the Edinburgh zoologists Turner and Allman ask him to come along to Dundee to the British meeting of scientists and give a first report about his research and theses. »I will give here my first lecture – and on top of it in English.« His participation brought Dohrn many friendly relationships. Even in England, Darwinism was still much debated, but the energy was appreciated with which German science – in contrast to France and Italy and particularly among the young – had received Darwin's questioning.

Above everyone else, Thomas H. Huxley,[16] who was familiar with German language and thinking, valued him highly. He drew Dohrn close to himself and to the »house rich in daughters.« The German became a merry companion. He wrote: »I read today some chapters of Mill's Utilitarianism and found the word happiness several times. If I had to define this much-wooed word, I would say: ›Go and see the Huxley family in Swanage.‹ A serene feeling of happiness always touched his memory when he thought of their common days. Huxley, 14 years older than his new-won friend, played a role in England at that time as interpreter and fighter of Darwin's doctrine, similar to that of Haeckel in Germany, although he held an incomparably stronger position in the spiritual and social consciousness of the nation. He shared with Dohrn not only scientific but also general interests. The correspondence reports on goals and disappointments in scientific work, but it also touches on political thought, because Dohrn could be sure of a similar basic mood in questions of international understanding, of the woman's role, and of social evolution. Conversations about Goethe during these visits found a long echo and Dohrn told of how he had bought and systematically studied Goethe's entire correspondence – Huxley should please not omit »this profound and incomparable source of humanitarianism.« Then again, after a renewed reading of Macbeth, the following characteristic sentences: »I wish I could communicate with a great actor – what different dimensions the penetration into such a tragedy would receive. You know, the only German poet who comes near to Shakespeare's tragedies is Heinrich von Kleist. I re-read him, too, recently, and must remind you again of his three or four main works . . .« Huxley should please advise him whether anyone in England would publish the *Prince of Homburg* »even without pay.« His Scottish friend Grant had a translation lying on his desk: »It is good, I guarantee, I who know the work almost by heart.«

Huxley's remark in September, 1867, »Thunder and lightning!« had not been a total agreement – a letter of Anton to his father states that his theses had first encountered only an amused surprise. Still, the encouragement to secure the foundation was great. This shows throughout the correspondence between son and father in those months: Carl August was in good humor, and was willing to finance a trip to America, where the development of giant crayfish was to be studied; his son would take care of occasional entomological findings.

He had been away from Jena since spring, and amidst his English impressions he almost feels a little homesick. »I have a little longing for my room among my books, and the simplicity of Jena life.«

The stay in the foreign country had sharpened his outlook. »I must admit,« he wrote to the Stahrs, »that my respect for German education is constantly growing. Since I have seen the English universities, I understand why there are more nonprofessional scientists in England than in Germany, and why even such men as Mill, Darwin, and Lyell are not professors. English students are more like high school students compared to our students who, despite their dueling, color-carrying and beer-drinking, are much more informed and educated.«

He could be certain that in Jena there were people who were looking forward to his return in the fall of 1867. Something like a circle had been formed, composed of very different disciplines. There were men among them who would attain great scientific fame. Others remained only the acquaintances of his youth. He reported on the evenings with the philosopher Kuno Fischer. Sometimes he mocks »the very kind Ebers whose thinking is somewhat lame: he cannot get beyond the dear Lord, although He does not cause him sleepless nights.« His closest friend was »a Strauss junior, by the name of Overbeck.« It seemed strange to him that this Overbeck, »who shares my convictions and is a unique, clear-minded wonderful person«, is a theologian. Franz Overbeck left the university the same year as Dohrn; he won fame through his work in Basel and his relationship with Nietzsche.

Already in the fall of 1866 his friend Hildebrand was beginning to appear as »the much more important son of his professorial father«. Adolf Hildebrand was at that time 19 years old, a young man of a luminous gaity. He was just preparing to go to Rome for the first time, where he was to meet Hans von Marées, who became a fateful figure in his first development. Dohrn knew him through his father Bruno Hildebrand, who was a liberal economist and former fighter of 1848. Young Hildebrand had been a student in Jena at the private Stoy School, where there were a few interesting teachers. With one of them, a young Scotsman, he became friends; the bust he made of him opened his way to sculpting; this was Charles Grant. [17] Grant taught English there, and gave lectures on English literature; he published a small critical study on its past hundred years. He also learned to love German intellectual life. He had little money and his situation and decisions were insecure. Dohrn took care, almost violently, of the somewhat younger man; he prevailed on God knows whom to find connections for this quiet and inexperienced man and finally tied Grant's fate completely to his own.

Another man also taught at Stoy to fight his poverty with the meager extra income * gained as lecturer in physics, Ernst Abbe; [18] he is mentioned for the first time in a letter of January 25, 1867: »Tonight I was at Dr. Abbe's for two hours, a mathematician who is beginning to give me satisfaction because he is so superior to me, particularly in philosophy and in a collected calm . . .« The two young men who met there might well seem like opposites by temperament and birth. But they fostered their friendship by common meals and eager walks – Abbe's helpful advisor and later father-in-law, the educated, very mild Professor Snell, was often the third member of their discussions. When he looked at his friend, Dohrn could appear to himself as a darling of fate, roads were smoothed for him, the world was open. Abbe, on the other hand, the son of proletarians from Eisenach, had fought his way to science by strictest diligence and under severe privations. It is this moral energy which made the strongest impression on the sensitive Dohrn (»unselfish, great and noble, one of the most profound thinkers I know«). From this youthful friendship onward, Ernst Abbe remained for him the measure of true humanity. Abbe was only the promise of future achievement at that time. The connection with the optical workshop of Carl Zeiss, from which his epoch-making achievements concerning the knowledge of light waves and the technique of the microscope were to be derived, did not yet exist in 1867. The boldest imagination could not have recognized in the modest man the later producer of a large industrial organization of a very particular social structure. The two might have agreed on the basic political development, because there existed in Abbe, inherited from his father, a humanitarian direction like that of 1848. But, unromantic as Abbe was, he was less impressed by Lassalle than the young Dohrn. The latter told later that he, stemming from the »bourgeoisie,« (little as such a

* Beginning university teachers had no salary. They received only students' fees.

cliché corresponded to the atmosphere around Carl August), had introduced the contemporary worker's son to presentday socialist literature. The friendship of the two men – even though there was at times little contact because both were obsessed with their work – lasted until Abbe's death in 1905 and had acquired new warmth in the last decade – they served each other in their work. Abbe remarked in 1896 how important Dohrn's enterprise in Naples was for the development of his work. And Dohrn knew how much the exact biological experiments owed to Abbe's microscope. But this was a later experience. What he owed to young Abbe was the example of a well-controlled energy of thought and a brave attitude, even in failure. This was soon enough proven.

Gegenbaur and Haeckel did not go along with Dohrn's hypothesis – the proof of which he believed to have found in England – but rather waited skeptically; at first this did not bother Dohrn too much. He was glad that the human situation with Haeckel had been worked out, that the »sanctimoniousness«, which he had criticized at times, had disappeared. He was very proud when on November 26, 1867 Charles Darwin sent him a very warm letter on his lecture at Dundee. »This I call«, he wrote to Stahr to whom he had sent a copy, »my scientific knighthood dubbing. What lends this letter particular value is that Darwin wrote it himself – something he rarely does.« Stahr was afraid that in his joy Dohrn had omitted to read Darwin's reservations. His answer tried to correct this, and is characteristic of the sensible way in which Stahr took care of his educational task. »I know your nature and your inclination to have exaggerated expectations where you feel favored by fate, and on the other hand to become despondent when the opposite occurs.« Therefore, he warned: »If you succeed in going forward to what your great master calls ›a triumph in science‹ – that would indeed be a great success. In the meantime, I beg of you, my dear Anton, not to overlook the fact that Darwin's judgment about your alleged discovery is only hypothetical. He writes, ›If you finally succeed,‹ he does not yet think that it is definite. Darwin still holds on to his opinion. And if Darwin could be wrong, you will also have to take into account the possibility that you were wrong.« This warning was not without consequence. Dohrn had registered for his habilitation on November 11, 1867. Haeckel gave the faculty his opinion on the offered material on December 5: »The author displays (in his embryological work) an estimable talent, both for investigation and observation, as well as for judgment and presentation. The work is, from all these viewpoints, equal to the best embryological investigations of our time.« A second study addresses itself more to the realm of »zoological speculation«. Haeckel remarks, with regard to the manner of the author, that he seems »not to proceed with the necessary caution and critical thought«. He had come to know Dohrn as an assistant. What he still needs are »diligence and persistence . . . At any rate, his disposition and talent are excellent and his education and interests are many-sided. But precisely this circumstance led him in former years to underestimate the difficulties.«

Not until January 25, did the »satisfactory« defense of zoology and philosophy take place – it had been postponed several times. For, in the midst of Dohrn's initial insistence on an early term, a temptation of a very different kind had appeared, one which mightily stirred his imagination. The Edinburgh zoologist, Turner, in whose house he had been received as a friend, offered him, in the name of the Indian Government, an unexpected opportunity to take over the directorship of a newly founded natural history museum in Calcutta. He was offered 4000 to 5000 Thaler. That would mean complete independence from his father and also scientific and organizational tasks which were not pressed into too narrow a frame. He reported the offer to Huxley who, having traveled for years as a ship's surgeon, knew the Indian world. Huxley advised urgently against this work, which would demand a lot of bureaucratic responsibilities, and even more against the climate, which would kill him within half a year. Also, he found Dohrn too valuable for such a position. »We have a saying that it is unwise to use razor blades in order to cut rocks.« (January 15, 1868). The skepticism with regard to health seemed justified. A serious sickness – of all times, in the late winter of 1867–68 – made him incapable of finishing his projected work. A trip in spring took him to Weismann in Freiburg, to the Stahrs and Carl Vogt at Lake Geneva. It also took him to Winterthur, where he finally met Friedrich Albert Lange after almost 2 years of planning. »I liked Lange very much. We talked a lot about social problems. I did not tell him much about my own work – I was thinking of your warning.« The weeks in Montreux did him »a lot of good«. Stahr probably repeated his educational warning against premature self-assurance.

His inaugural lecture took place early in May. He had chosen the topic *Kant's Relationship to the Theory of Descendance.* These were the two intellectual problems which had occupied him most during the past 2 years. The study which he presented (it is preserved in manuscript form) is free of any enthusiastic exaggeration of the answer, a very neat investigation of Kant's opinions on the methods and tasks of natural science, done with a certain philological accuracy. He did not, of course, suppress some personal convictions which demanded a revision of the relationship between philosophy and science. He talked of the period when »scientists spoke with particular satisfaction of »Hegel's nonsense, Schelling's chitchat, Fichte's craziness«,* and when every scientific Liliputian, with his »exact observations«; believed himself to be a god compared to »their fantastic natural philosophic speculations«. Kant had been spared at that time, or else he might even, as an »unknown god«, have been offered sacrifices. By investigating Kant's work and placing him within the contemporary situation of natural research, he found that Kant was »dogmatic«. Kant adopted for the existence of »races« the maxim of some preformed germs and natural dispositions, and therefore was opposed to the idea of evolution. But he kept the teleological elements of his doctrine (which

* These three German Romantic philosophers were concerned mainly with the philosophy of nature rather than with scientific research.

assumes the purposefulness of nature as a special concept a priori *) free of crude anthropomorphism. This is presented in subtle observations and proofs. And Dohrn illuminated the contradictions in the evaluations of the mechanic and the teleological viewpoint in the realm of organic nature. For the disciple of Darwin's doctrine (who simultaneously owes his freedom from materialism to Kant – through F. A. Lange) the result of specialized investigation is almost disappointing. But he found in the position of the philosopher toward natural science a decisive and fertile beginning, namely when Kant, turning against G. Forster, demanded natural history instead of description of nature. It is the *historical* viewpoint which necessarily regulated the power to overcome a merely teleological conception. The faculty was convinced of Dohrn's ability to teach. The dean noted »that the lecture, although keeping somewhat away from the discipline of zoology, was very good in form and content.«[19]

Dohrn taught a class on arthropods. In the middle of May, he wrote to Stahr: »My lectures go quite well. I have between five and eight students; some are coming quite regularly and are really interested in the analysis I present. I speak quite freely and have not yet felt a lack of material in my head.« A young Baltic student was attending, whom he »rather feared than wished there«, because he was already a good zoologist and, at some other place, had spoken »quite positively« about him. This was Nikolaus Kleinenberg, whose fate was later closely united with his.

The secure tone of his academic start broke up suddenly: the foundation of the genealogical order, on which he had worked so long, did not stand up against closer analysis: »since then I could not find the strength to write,« he wrote in a letter of May 29, 1868, »because I have been hit by very bad luck. In my book, a funda-

* A priori: an expression mainly used by Kant. It means knowledge without experience, in contrast to »a posteriori«, which is knowledge derived from experience.

mental tenet is based on wrong observation, so that all my work is in vain. If I do not want to expose myself to the worst criticism and scientific ruin, I will have to hold the book back and destroy it. I am very disturbed, and can barely stand. It was a terrifying downfall!« A few days later, he recovered himself: ». . . if I may trust my returning feelings, this blow will not kill my courage to fight the demons of my all too sanguine nature. I will turn this bitter experience to my advantage. I will think and investigate more slowly. Please do not think that the reason for my hope is levity. By God, I spent two awfully bad weeks, and what I always suspected has been confirmed again: the irritation of my nerves due to a bodily state tending toward the pathological. Today is the first day on which I feel more free . . . I will probably go to Scotland again and take up, and deepen, my investigations at the same place where I worked once before so intensely . . .« (June 4, 1868).

In this self-analysis, we hear for the first time the motif that was to accompany the rich life of Anton Dohrn so tragically: it was and remained endangered by breakdowns of this kind. In this early case the depression was caused by the disappointment which followed the high tension of self-assurance. Later, it appeared sometimes in the middle of relaxed work – if one can speak of relaxation in such a tense nature. For weeks, sometimes for months, he was almost paralyzed, tired, irritable, although he tried to hide his nervous condition behind his polite, humane manner. The difficulties of the enterprises he had started grew like giants before his consciousness and weighed on his mind, until the clouds disappeared. He thought that this bad situation was inherited from his mother's side, in whose family cases of depression had occurred. His mother tended to take life very hard, although not with the same tension and wild span between an enthusiastic will to win and the stupor of feeling annihilated. In later years, this condition eased, but the threat was always present. We have to keep in mind this mental disposition in order to understand some violent curves in Dohrn's behavior. The goals for which he had to fight fate and his ability to tear himself away from depression are all the more admirable.

It was a hard blow that he had given in to »his suicidal temperament« and his »rashness« had deprived him to a large extent of the joy over his project. Nevertheless, the subject matter had captured his imagination, Scotland would be of help, the work would have to be repeated. After the sojourn in Millport he would continue his studies during the winter in Messina. Although the hard work during the summer of 1867 had been in vain, he wanted to protect himself against the reproach that his imagination could not be regulated by thoroughness. Stahr did not quite trust the Messina plan. Dohrn defended himself against Stahr's remark that »this was just one of those ideas« by saying that »he had to defend his name as an observer against the damage which I did to myself . . .« His father, satisfied with the »habilitation«, agreed to the plan. He did not yet quite understand the crisis in the scientific thesis of his son. He was only sorry that Anton was obviously moving away from the realm of entomology. It had been part of the father's pride that the *Entomologische Zeitung* could print contributions of the growing scientist – which now did not materialize. The answer to an appeal (only in June 1869) is very characteristic, also from a scientific point of view. It describes the entire situation in which his work of 1867 and 1868 had been caught: »Certainly a more versatile work – crustaceans today, spiders tomorrow, and insects the day after, then annelids, etc. – would be more beneficial to me personally, with respect to my University career. By limiting myself completely to a single scientific project, I deprived myself more or less of the opportunity for a universal zoological knowledge. But the way things are today, zoology requires wearing blinders in just this way. Darwin has overthrown the old systematic framework of beliefs, and we stumble among the wreckage with more or less awareness. One style has perished, and a new style is supposed to be built. My inclination and habit to occupy myself with the arthropods is strong enough to satisfy my zoological needs. I am never bored dealing with this class over and over again. The problem, however, is so great that many more years will pass before it can be considered as solved in broad outline. I have become more and more convinced that the current opinion to combine crustaceans, spiders, millipedes, and insects in a genealogical unit like fishes, amphibians, reptiles, birds, and mammals, has to be abandoned. I had tried to prove the opposite, that is, the unity of the four arthropod classes, and all my initial publications maintained this position. Now, I must refute myself . . . If I have to formulate the result precisely, then I state: ›I, Anton Dohrn, see in my embryological investigations no way by which a connection of the embryology of arthropods and insects (or spiders) is possible . . .‹« And to the extent that the letter considers other possible relationships (Dohrn is still involved in crustacean embryology), he states: »It would take a whole life to fulfill this task adequately. Failure is not even included.«

This sounds very different from the brisk certitude of the conqueror in the earlier statements. The goal is not anticipated in a hypothesis, rather, the road leading to it is prepared on a long-term basis. But the hypothesis is not dismissed altogether. Research cannot dispense with a heuristic principle, and Dohrn's own scientific work remains connected, with stubborn persistance, in certain theses concerning the evolutionary relations of the vertebrates. Through decades there are continual new starts to collect evidence that is arranged in the context of a basic concept. But it is the experience of these years which liberated him from dogmatism, or at least protected him from hardening, in view of the fluctuations and changes of biological tenets.

While Darwin remained the one who had awakened him, and while he tried to comprehend the revolutionary effect of the English thinker on the fields of social economics, psychology, and education, he pays homage at the same time to the baltic scholar, Karl Ernst von Baer, whom one might consider as the counter-figure to Darwin in the interpretation of the fundamental biological laws. In the discussion sparked by Darwin, he remains the classical representative of a teleological view of organic development, but at the same time the promoter of exact observation and of the unraveling of the elementary natural phenomena.

6 MESSINA

The second trip to the Scottish west coast, at the end of July, 1868, was supposed to cure Dohrn of his depression. But that was not so simple. The atmosphere in the house of Mr. Robertson was free of scientific pretensions and when Dohrn, with his host, rowed among cliffs and had all sorts of adventures, he did not have to feel defeated. He limited himself to »a small work which was complete in itself and brought some new details«. It filled a gap in science, »even though a small one«. The letter to Grant, to whom he wrote this, is still in the shadow of a self-tormenting view: never again should a mirage make him forget his own situation. He reminded him of his last stay in Millport: »that was the most beautiful period of my life and also the only one in which I felt that I could have made something of myself. Now it is over – not that ideals have disappeared – no, the simple existence free of fear is gone,« since my »demon let me taste the most detestable of all deceptions«. And in telling his friend his further plans, he said that the trip through France should go fast: »nothing is more dangerous to me than to have spare time, because I will certainly fill it with gloom. Restless activity is the only weapon against my sickness.«

The concentration on his work was helpful. When, in the middle of September, he went for some weeks to the beach of Plymouth, in order to work there, too, on crustacean larvae, he confirmed his eagerness, which forbade him any entertainment in a more relaxed form: »I have done a good deal of work and it is not my fault, if the good name of German diligence should get lost abroad.« With Robertson, he had not only done research with the microscope, but also experimented with constructing equipment. They had made a transportable aquarium, which might be of useful service in the new environment of Sicily, because such pleasant working conditions as those in Millport would hardly be found there. The visit with Huxley gave him back his inner freedom. He had written earlier to his father, how much luckier he would have been if he had had this man rather than Haeckel as a teacher. Now they found themselves going back to Goethe – *Wilhelm Meister* and *Faust* accompanied him on this trip – it would have been better to be an assistant to Huxley than a professor in Jena.

In the middle of October, 1868, Dohrn arrived in Messina. He had rushed through France; the trip from Marseilles through the Tyrrhenian Sea was marvellous; when he passed Caprera, he took his hat off – that was meant for Garibaldi, whom he revered: »The rabble on the Seine river must be destroyed by this ancient character,« he had already written a year ago. This symbolic gesture greeted the political fate of Italy, in whose development (he did not know this yet) his own future would be included. The reception in the harbor was full of color and noise and made the newcomer serene and cheerful. But he impatiently awaited the freighter from England which was to bring the heavy and complicated apparatus which had been loaded up for him in Glasgow. Since the eighteenth century, zoologists had known and recognized that the Sicilian coasts were distinguished by a particular wealth of forms and species of marine fauna. The Italian Scientist, Spallanzani, had in 1788 made his experiments and observations and had studied in particular the pelagic fauna. He had been followed at the same place, two generations later, by the German, Krohn. For the interest of German research in marine fauna it had been decisive that Johannes Müller, the founder of modern physiology, after having worked in the northern seas, had himself indicated to his students the greater wealth of the Mediterranian coast, where Carl Vogt had tried to locate his various unsuccessful attempts of laboratory research projects. According to a joke of the Italian de Filippi (Todaro of Rome mentioned this in 1897 in his speech during the jubilee festivities for the Naples Station), Messina had become, after the middle of the century, »the Mecca of German private professors.«

There the »private professor« of Jena, Anton Dohrn, had traveled, but he brought with him unusual, not very pilgrim-like, formless, and over-heavy equipment. Later on he chatted, with indulgent self-irony, about the effort of having hauled to Sicily these cases which had been constructed in Scotland. But something important came of it. One can see in these makeshift, transportable aquariums the form of the later zoological station. He described the matter to the Stahrs as early as September: »An aquarium three feet long, two feet wide and one foot five inches high travels by steamer from Glasgow directly to Messina. The floor and the two narrow walls consist of slate, the two large walls of mirror-glass. I myself fit the whole thing together in Messina with the help of screws and a special putty. To produce a continuous stream in the aquarium, I have two more cases, equally large, built of slate. One of them stands under the aquarium, the other next to it or above it . . .«

The idea made sense. For the work of the scientists, on any beach, always meant a somewhat adventurous improvisation. There are rueful and comic reports of how in a hotel room or somewhere else a scientific workshop was produced, with borrowed pails of water, of fights with porters and hotel owners. One could do some anatomical work, also make some morphological observations, but all biological observation remained doubtful. Through the aquarium Dohrn found his hopes »completely fulfilled. Already in the first days an important question was decided, namely whether or not certain transparent, paper-thin, flat long-legged crayfish were the larvae of spiny lobsters. The latest scientists had negated the question, the aquarium had confirmed it irrevocably.«

Stahr was informed in detail that all these eggs would die »unless they are kept in an uninterrupted flow of sea water«. He had become acquainted with the importance of moving and renewing the sea water during his studies in Hamburg a year and a half ago. Now, full of enthusiasm, he invented easily transportable aquariums made of wooden cases faced with zinc and glass walls, since his slate cases were too inconvenient. »Every scientist who goes to sea can take his aquariums along.« This would be of »boundless use«.

It had become important for the optimistic and happy mood with which the new environment endowed him that he should quickly find friends. He was not made for solitude, with his lively need for exchanging ideas: he needed the echo, the intimate contact and intellectual exchange, particularly in this environment where the power of the landscape lured a person, and the richness of history oppressed him. By accident a young Russian scientist, Miclucho-Maclay, whom he had known in Jena, was also working in Messina on the brains of fish. With him he undertook some excursions. The attempt together to scale Mount Etna on January 7, 1869, almost ended for Dohrn in a catastrophe. The Sicilian guides refused to go along on the actual climb; with a certain obstinacy, but insufficiently equipped, the two men tried to force their way to the top. His »bad luck«, as he reported to Huxley, seduced Dohrn to climb up dangerous rocks in order to shorten the way through the ashes. He lost his way, fell, and slid 400 to 500 feet on an icefield, keenly aware of the fact that he might break his head on one of the rocks. Amazingly enough all went well. He lost his watch, his clothes were torn, there were contusions over all his body, face, and hands, he had bloody wounds, also a dangerous tear in his kidney – it was a wretched painful way down.

When they finally reached the bottom he was loaded onto a carriage »like a piece of wood« and taken to the train. Work was spoiled for a few weeks until the wounds were healed. But it so happened that Carl August Dohrn was once again in Rome with his English friend, Stainton. So Anton went to see his father. The latter was happy to spread out before his son the wealth of his knowledge and old experiences. And the son was grateful for the friendly openness of his father in this overwhelming environment. He had not seen him for a very long time.

Miclucho-Maclay took Dohrn to a Russian family whom an unusual fate had driven to Messina a few years earlier. This relationship, which soon acquired the character of a close friendship, was to become important for Dohrn's work and fateful for his life. The Councillor Georg Iwanowitch von Baranowski was 47 years old; he had had a successful career in the inner Russian administration, which suddenly broke off in 1863. He came from a wealthy family that had settled in White Russia in the fifteenth century. He was the oldest of a numerous family and had been sent to the newly founded law school in St. Petersburg in 1840. There he excelled in diligence and talent. After he had been in the administration of the Baltic provinces, he became governor in Ufa and Saratow: »At that time Alexander II was preparing the liberation of the peasants. Baranowski, who was a friend of Alexander Herzen had strongly participated in this work together with his younger brother, Alexander. The administrative situation changed with the outbreak of the Polish revolution, which violently divided the family with regard to sympathy and position taken. His mother was Polish and the sisters and brothers, differing in their political positions, fought on opposite fronts. Even more strangely, the same thing happened in his wife's family (he had married Katharina von Timler in 1855), almost to the point of tragedy. The Timler family, coming from Mecklenburg*, had served in Russian military service. Katharina's father had married a Polish woman – this family, too, broke up. One Timler was banished to Siberia and, having been pardoned, finally ended up in reduced circumstance as a greaser on the railroad (in that work Dohrn met him). Another brother was given to the Battenberg's Alexander as adjutant and controller in Sofia when the Bulgarian experiment was made. He moved up to the rank of lieutenant general. Katharina, the wife of the governor of Saratow, a beautiful and proud woman, passionately adopted the Polish cause, not only in what she said but also in helping the captured insurgents who had been transported into her husband's department. The conflict was unavoidable and dangerous for Baranowski, because he, being a very fair man, had made enemies among important men in the Department of the Interior. He answered a reproach made to him on account of his wife's behavior with the request to be discharged from his position and to leave his country. First, he went to Geneva in partial emigration. The Rus-

* Mecklenburg, a Prussian province.

sian Black Sea steamship company offered him a position as their representative in Messina. He occupied this post when Dohrn entered his circle in 1868.

Messina was intended to be only an intermediate position for this active and circumspective man. After the opening of the Suez canal he traveled in East Asia for this company to investigate the possibility of expanding the direct Russian sea traffic. Thus, his broader economic activities started, which were to take him back to his Russian homeland within a few years. Dohrn, with his lively interest in everything unusual, was fascinated by the fate of the high Russian official who had changed from a Councillor to an agent of navigation. He was also fascinated by, and attracted to, the lively atmosphere of the house. This was in the first place political and literary. But the family opened up to his scientific interests. He enjoyed teaching the children of the house (Maria, age 12, and Helene, age 10), both very lively: he took them on excursions and let them look through the microscope. This was an educational pleasure and a happy relaxation. For the girls he became an admired mentor.

Dohrn was satisfied with the result of his scientific work in Messina. He succeeded in making the eggs of a palinurida slip out into the flowing sea water. For this, he had to thank the cases he had brought from Scotland. In his conversations with Baranowski, a plan came up to make all of his equipment useful for later scientists. The Swedish consul took over the preservation of these objects. There, Dohrn also left for his successors a kind of diary and practical advice with scientific observations and practical ideas. It was supposed to make their work easier and more certain. He put a motto in front of it: »Vestigia non terrent, sed adjuvant« (Traces are not frightening, but helpful). Looking back later, he said that, »From this moment on, my thinking remained no longer free of the thought of creating a little settlement in Messina which would be called ›Zoological Station‹ and would, as a kind of free possession of science, be open to everybody who went to Messina to work there.«

7 IN THE MIDST OF DECISIONS

In the Spring of 1869, it was hard enough for Dohrn to »leave Messina, where gradually also the human element began to hold me.« The letters to Stahr emphasize how useful Italy had been for him. »It made me much calmer and let me renounce having too many enterprises.« The return to Jena was not as easy this time as on his return from his first Scottish stay. Grant was trying work in Berlin as a journalist . . .

From his lectures only one practicum materialized. It seemed as if the fear, expressed in 1866, was justified and the smaller universities had to expect a reverse development. In those 3 years Jena lost 200 students. Anton reported to his father that Haeckel had 30 students instead of 200 in his sensational lecture and only 5 assistants instead of 12 to 14: »In this situation I am glad to have caught two students for my entomological teaching cramps . . .« He is somewhat scared that »among the Darwinists there is a burning conflict«, Virchow against Vogt and Haeckel, Schleiden against Virchow, etc. The scientific exchange with the two professors had a certain liveliness, but the intimacy with Haeckel was lost; it never existed with Gegenbaur. In a letter to his father he characterized the exchange with the latter: »In the first place, he is very violent, in the second he prefers to be the only talker. And when he has expressed his opinion for a long time, he believes that it was a discussion, in which his opponent lost without having been able to say a single word.« In October he reported to Hökendorf that Gegenbaur, speaking to Czermak, had »highly praised« his essays published in the Jena Journal. He thinks this caused his lecture of the Winter semester (comparative history of the development of arthropods and ringworms) to »show twelve listeners instead of the expected three or four« – this number then rose to eighteen. Only two came to the *General Entomology*. »The students seemed to have found out that the rabbit is buried in the theory of evolution, rather than in systematics.«

In the following years (1870–71), eleven essays appeared on *Research on the Structure and Development of Arthropods*. They were the result of investigations on numerous kinds of sea crayfish in Scotland and in Messina and on fresh water crayfish. In these essays hypothesis gave way to exact observation, and the *History of the Crayfish Family*, published in 1870, looked very different from the all too bold summary of 1867. Although a few bones of contention with Claus and others remained, Dohrn was now recognized as one of the foremost experts in crayfish morphology. His new concept, »quite revolutionary« in comparison with the ruling doctrine, that arthropods don't form a genealogical unity, but that crayfish and insects »developed separately from the annelids« became prevalent. He described to his father the state of his concept on August 29, 1869: ». . . I have learned from the mole cricket ten times more than I would have expected. Thus, the theory of the separation of crustaceans and insects is not in the least shaken. On the contrary, it seems to clarify everything so well that it can never be buried again. But at the same time, the ties between annelids and insects, and through them to the vertebrates, become stronger. The embryonic processes of the latter, to the extent that they must be related to those of the annelids, since they descend from annelids, are completely identical with those of insects. In this way, I connect up with Rathke and old Baer, whose successors in insect embryology did not want to hear their opinions. Weismann's investigations of flies are undoubtedly the most brilliant work that embryology has seen since Remak's developmental history of the chick – but he failed in everything that was related to the most fundamental processes, obviously because his object was too unfavorable for investigation. I have been able to clarify these most fundamental phenomena, and in this way I have disposed of the exceptional position of the insects. Everything proceeds according to the theoretical postulate – and the practitioners of the microscope were up to then in error.«

After he himself had limited the hypothesis and theoretical claim of 1866 and 1867, his tone had become self-assured once again. At the same time he put on paper the lecture he had promised to give in Jena in the fall on »Darwin's theory and the reform of psychology.« (»I scribble with great delight.«) In this lecture, Darwin's theory is not seen within the frame of zoology: »It is *an idea*, the idea of development, and it is a *principle*, the principle of natural selection. This is completely new and an unrestricted property of Darwin, the former as old as the intellectual world, as old as all philosophy. But it has been suddenly placed at the *top* of intellectual thinking (from a subordinate position) and for this reason is so new. Darwin's theory is treated with a suggestive as well as confessing force which places him among the Great Men: Copernicus, Galileo, Kepler, Newton. It is treated not only as an heuristic means which points at the understanding of the development of organisms, at the clarification of the laws of heredity and adaptation; it is rather set up as a goal »which lifts, with the same weapons of the mind, the seemingly impenetrable veil of the future, toward which mankind moves«. This is the optimism of a generation which knows that it stands at the crossroads of scientific questioning. Precisely a nature like Dohrn's, with his universalistic direction, felt the desire constantly to break up the scientific limits and to force the individual scientific findings into a totality, which then would be followed by practical considerations. Such a strong impetus also knows resignation and the relativity of the value of individual findings. It is a curve with uneven shifts. Now that he wants to see psychology faced with new tasks and a new method of observation and of collection of material, he reaches out into the work of modern linguistics, in which the elements of evolution have already become productive. Max Müller is an example. From there it is only a step to the animal – and child – observations of Alexander Bain. His experimental psychology interested Dohrn very much in those years. It is important to him to find the direction for a new doctrine of education. Here »we are still stuck in the Middle Ages, in barbarism.« It is not enough to »boast of enlightenment with self-complacency. To us enlightenment consists less in having learned so much than it does in the fact that we are more enlightened every day in what we still have to learn . . .« This insight contrasts with the violent, almost pathetic statement of how one may transform educational theory in the family, in school, in public life, by accepting Darwin's insights. Yet he clearly saw the danger that threatened: »Since Darwin's theory became popular« it has also made the theory of hereditary shortcomings popular; this could become a danger to the seriousness of education. For many failures can now be dressed in the mantle of scientific propriety. To oppose this one will have to emphasize and develop the fundamental statement of Darwin's theory: namely adaptation. This was developed in several examples which show how Dohrn himself had practiced the observation of children and teenagers. He described how one achieves typical insights by collecting materials of observation and how much of the latter exists in the collections of letters and in autobiographies.

The great lecture had, within Dohrn's work, only a minor importance. But it is nevertheless essential for characterizing his basic mood of that time, perhaps even the peak of his general social eagerness for reform, his desire to be efficacious beyond the merely scientific sphere. The expansion into social and political problems of his time was to accompany him throughout his life. In the future he avoided taking public positions, and also avoided the polemical fights against the claims of the church for school education.

In the summer months of 1869 Dohrn prepared himself for his participation in the German meeting of natural scientists in Innsbruck. »There I will give one lecture after another.« His father, who visited these meetings quite regularly, was also in Innsbruck. In the *Entomologische Zeitung* he reported about the lectures and various discussions of his son. The program mentioned four lectures: »The fundamental phenomena of insect development and their meaning for taxonomy«; »Homologies of Organ Formation Between Articulates and Vertebrates«; »The Taxonomic Position of Prycnogonides«; and »On the Supposed Genealogical Connection between Crustaceans and Tracheates.«

He was satisfied with the reception. »The meeting«, he wrote to Stahr, »was extremely useful. All authorities expressed their recognition in the presence of my father. He himself is finally convinced – now I am free.« He met Helmholtz very informally. He also heard in Innsbruck that he had been on the list of candidates for the professorship in Würzburg.

From Innsbruck he traveled with Carl Vogt to Venice, which he did not yet know. Being with Vogt may not have been unimportant for the clarification of the thoughts which had occupied him since he left Messina, because Vogt was his predecessor in various experiments. As a student, Dohrn had greatly appreciated the works of Vogt: »Vogt is a genius,« he had written in a letter to Haeckel in 1864, but he no longer held that opinion. Under the influence of Lange he had clearly separated himself from Vogt's determined materialism. But during his visit in Montreux he had come into closer personal contact with Vogt himself, who was substitute professor of zoology in Geneva, and he liked the good-natured and always helpful man, although he was a little blustering. At the age of 31, Vogt (at that time a professor in Giessen) had become a member of the Paul's Church, * a brave orator of the left. Fate made him become, in the weeks of the Stuttgart parliament, one of the three »Reich's governors« and that meant the end of his further academic career in Germany. Because of his kind care he was considered for a while as the head of the German emigration of Switzerland. But he soon sought and found the contact with scientific teaching that he needed. He also took part in the political life of Switzerland as a National Councillor. With him, Dohrn could discuss his research in marine fauna. Already in 1844 Vogt had been busy as adviser of a French enterprise which intended to create an observation post on a coral island. In his first years in exile, in 1851, he worked in Nice on such a plan. In 1863 he also worked with his

* Paul's Church. A church in Frankfurt where the leaders of the 1848 revolution met.

friend, the Italian Secretary of Education, Matteucci, on the foundation of »a kind of zoological seminary for teachers of natural sciences« – the matter seemed certain, when the Secretary resigned. Vogt may have told his young friend about it. When, a few years later on July 26, 1872, he presented, in answer to Dohrn's question, the various efforts and their course, he could also report about the latest attempt to create such a station, in January 1872, near Trieste in a side building of the Miramare Castle. The Governor supported him; Darwin's recommendation of the plans was expected to help in Vienna at the ministry of education – here, too, a change in personnel destroyed their hopes. »As you can see, this is a story of failed efforts which I do not want to enlarge by quoting some essays . . .« Dohrn always acknowledged Vogt's idealistic precursorship – the original »ancestor of the plan« he wrote in a letter to Werner von Siemens in 1880. When he was in need he could always count on Vogt's advice, his sincere warning, even on financial aid from this man without funds. Vogt's diligent pen was more than once at the disposal of the propaganda. That was basically welcome, but not always without danger, for example, when the popularizing newspaper writer in him won out and Dohrn had to correct the impression, particularly to a man like Siemens.

These were worries of later years, not important enough to weaken, even in anger, the grateful feeling he had toward this jovial and efficient man. Always, for example, he defended him toward his father, to whom the name smacked of barricades, and praised the hidden aristocrat in this »Falstaff.« Now, in Venice, he felt well in this peaceful atmosphere. Somewhat adventurously, he interrupted their stay on the Lido, when the newspaper noted that Etna would shortly erupt. He wanted to experience this and left immediately. In vain did he wait 2 days, but he was now able to enjoy the beauty of the landscape of which his accident in January had cheated him. It was a somewhat abrupt decision. His return trip led him past Etna to the seaside resort and the Baranowskis. One feels that this renewed meeting was as important to him as the hope of witnessing the phenomena of nature.

III THE DARING EXPERIMENT

1 THE GROWTH OF THE NAPLES PLAN

For over three and a half years, Anton Dohrn had avoided visiting Stettin and Hökendorf. The reason was the story of his broken engagement. He wrote to Lange in the fall of 1866 about it and its effect: »As a sad consequence of a wrong education, I got engaged early and later dissolved that completely unsatisfactory relationship. Everybody resented this step, and since I had already earlier hurt the ›educated people‹ in my home town by sharp articles against medical charlatanism and the cult of homeopathic ›priests‹, everybody has now outlawed me morally and crosses himself three times when my name is mentioned.« This may be a slight dramatization, but for a long time the memory of the girl remained uncomfortable for him. But she had married in the meantime and had a child . . .

If ever his »reputation« had been as questionable as he described it, it was now forgotten. He was asked to give a public lecture at the museum started by his brother Heinrich. *Natural Science and Tradition* – its purpose was to keep the Biblical and the scientific world view apart and to reject, on both sides, the dogmatic transgressions of limits between religion and science. His appearance was followed by a gentlemanly discussion in the newspaper. He could be certain that he had regained his reputation among his fellow citizens.

More important than this was the discussion with his father. It is possible that Carl August Dohrn, if he was willing to listen, received an inkling of his son's plans. Already, in a very affectionate »thank you« note from Messina, in which he remembered their weeks together in Rome, there had been (February 15, 1869) a diplomatically veiled indication: ». . . When I breathe again in Jena the narrow Thuringian ›informality‹, Rome, like a rising sun, will join me and show me in bright light Jena as a *means to an end – Italy*. If you enjoyed showing me Rome, please accept as thanks the assurance that since then a sweet poison resides in my heart, which will barely be without effect in the future.«

Carl August had himself been an indefatigable traveler in Italy and carried in his own heart the »sweet poison«. In his youthful conflict with his own father, he had wrestled his rich years on the Mediterranean from him. He anticipated that his son would come to him with a demanding request, which could not come as a complete surprise to him. When his two other sons had reached the age of 30, he had given each the sum of 30000 Thaler from his inherited assets. Anton asked him to do the same for him* and suggested that he wanted to use the money for a zoological station, perhaps even for several. He believed he had acted very cleverly by adding this conversation, on the side, to a comfortable game of billiards, which always left his father in good humor. But this was precisely the wrong approach. It made his father suspicious; their two temperaments came into conflict. The conversation, so diplomatically prepared, ended in a quarrel, with a blunt »no« by Carl August and a defiant insistence by Anton that he would carry on the plan on his own. How to do this was probably not much clearer to him than it had been to Carl August when, almost 40 years earlier, he had left Stettin to create for himself something of a profession, even that of a soldier for Dom Pedro in Brazil. At that time, Alexander von Humboldt had appeared on the stage of the family drama as a deus ex machina. Would there be a similar figure in the fate of the son?

He reported to the friends in Messina on the situation (December 31): »I have completely broken with my father. He did not want me to settle in Italy and, in order to prevent this, he refused to give me the money. I declared that I would nevertheless go on my own account, without other resources than my head and my hands. I

* In December 1869, Anton Dohrn had his 29th birthday.

have connections with writers who will help me to publish what I write. I believe that I can live in Messina or some other place in Italy with 2000 francs a year.« He also said that he had just received a friendly letter from Darwin, and that he had told him and Karl Ernst von Baer of his ideas. In Messina, a small house and a boat had to be bought – he thought that all of this could be obtained for approximately 8000 francs. A collection among wealthy friends would help.

A few days later, these speculations were overtaken by surprising new events. In a sudden about-turn, the view of a new and practical approach seemed to open up. The story of this life-saving idea has often been told and has achieved an almost legendary character. As Dohrn writes: »Just after New Year's day, a few days after my break with my father, I went back to Jena from Stettin, where I had to be for my lectures. During the railroad trip my head was full of both the practical and the theoretical project. I almost felt like a martyr, since I was conscious of the fact that nothing was farther from my mind than wishing to go to Italy for pleasure's sake. I arrived at the Station of Apolda around 1 p.m. There was at that time no railroad to Jena from there, and the connection was made by means of a carriage or bus, the so-called »loafers«. I boarded a bus and, since the weather was clear and cold, I sat down next to the coachman. In Berlin I had had time to visit the rather new Aquarium whose director was Dr. Brehm, whom I already knew. As I sat in thought and the carriage slowly climbed the rising road, I suddenly had an idea. The Zoological Station could be realized if one built a large aquarium on the Mediterranean, the richest sea in Europe, and use its proceeds to cover the costs of a small laboratory. I immediately understood the wide implications of this thought, not only in general, but also for my present situation. I became so excited that I asked the coachman to stop and let me down. I wanted to walk to Jena on foot; I needed the physical effort to control my excitement, which became increasingly stronger. I also needed time and solitude in order to master and order my thoughts, which attacked me violently. I immediately left the highway and walked back to Jena by field and forest lanes. There was not a moment's doubt in my imagination, which worked plastically and quickly, that I had found the solution to a great task and simultaneously to my entire life's problems. The nervous depression which had held me prisoner for so long disappeared quickly and completely. I had a light-hearted and strong feeling of well-being which made me take great detours, so that I arrived in Jena only after four hours instead of the needed two.«

It was evident that for such a plan only a well-populated city, visited by foreigners, came into question: Naples rather than Messina. His friends Abbe and Kleinenberg, who he alerted immediately in the afternoon of that day, January 4, 1870, also realized this. In his emotional state he needed an echo and he found it: »All three of us were in that happy state of paradisiac innocence which usually forms the basis of a faith that moves mountains.« Perhaps it was lucky »that the two young men were as inexperienced in worldly matters as I was, and really had no idea of the foolhardiness my projects presented«. Looking back later, from the turn of the century, these thoughts also meant that: »Up to now much that I had done had gone wrong. My complete lack of knowledge of any business life and work could justify any distrust which people had for me.« At that time, he would not have trusted himself to overcome a tenth of the real difficulties which he was to encounter later. He did not yet see them. Looking back, he said that »only instinctive self-confidence and fearless recklessness would lead to success.«

At this point he believed he had found the philosopher's stone. Imagination works fast.

Stahr was informed about the plan on January 13, 1870: the project should be founded on securities. Two floors for the public, one for scientists, fewer grottoes than in Berlin, more marine fauna. Then the disarming innocence of the calculations: »I already figured that I could cover interest, management, and everything else, if I had 120 visitors per day for 9 months. And how many more will be coming! The beauty of the animals, the boredom of people, and when it rains! Congratulate me, the thought means cash, freedom, independence . . .«

A few months later, when he was already deeply into the bold and tough preliminary negotiations, he wrote a long letter to Thomas Huxley on April 24, 1870. It says, in addition to the progress of planning important things for the scientific program: »I hope that the aquarium in Naples will render a good income. I also hope for a good income from a small aquarium that I want to build in Venice« (from this he hoped to pay for two scientific assistants in Naples and one in Venice). »But I am going much further. The calculations I have done show that, after deducting *all* costs, I can count on a yearly gain of 1200 to 2000 pounds sterling. With this money, it will be possible to carry out my original idea of the Zoological Station . . . I intend to erect, for the time being, smaller stations in Messina, Gibraltar, the Isles of Cape Verdi, the Portuguese coast, and Ireland. Each such station requires a one-time expense of about 800 pounds sterling and a yearly support of 150 pounds sterling. Each of these stations should have a zoologist educated in modern zoology, a fisherman and boat, a sufficiently well-equipped aquarium, and a moderate library. All should remain subordinate to the central station in Naples, which will remain under my command, and the scientists of all stations must follow my scientific and other orders.

Thus, we will achieve a coordination of strength such as no science can display up to now . . . Once I proceed with the creation of such stations, England, France, and Russia, etc. will follow and we will arrive at a completely new form of biology. According to my inmost convictions, all of this is urgently necessary. Our knowledge of the geographical distribution of marine fauna is almost zero. Of the bathymetric distribution we know not only zero, but minus one. We need, as the most important item, a comparative physiology which can be derived only from zoology, not from human physiology. Embryology shows an immense growth. In other words, biology must become the science par excellence of the century under the pressure of Darwin's theory.

This I hope very definitely, if not to achieve, at least to further by the creation of the zoological stations . . .«

2 THE INVESTIGATIVE TRIP TO NAPLES

Ten years earlier, the Bourbon kingdom of Naples had been defeated by Garibaldi's bold expedition and had become part of the growing united states of Italy. There was no longer any question concerning the total political result of the enterprise. But the internal formation of the power play and the opinions were not ended. For the foreigner it had to remain largely obscure. The assimilation to the more rigorously shaped norms and habits of northern Italy was just beginning: the »two kingdoms« of southern Italy had competed with the Papal State in the doubtful question as to which of them was the worst with regard to police and administration. The dominion of the Bourbons had become more than ready for collapse, mainly because their last two representatives had worked with a stupid and vacillating policy. But traditions and interests had not collapsed at the same time. Garibaldi's victory had aroused the democratic elements of Sicily and Naples, which supported his short »Dictatorship« in Naples. That was already over. The folk hero of the national revolt lived in a kind of banishment or at least in a state of distrustful surveillance – the government of the Kingdom, which at that time resided in Florence, did not want his temperament prescribing the laws of further action. His followers, the »Patriots«, formed the opposition in Naples.

The city administration was moderately liberal, which corresponded to the still living regime inherited from Cavour.* It was difficult to estimate the number and strength of the Bourbon and clerical elements. With its 420 000 inhabitants, Naples was certainly the most populated city of Italy, but as far as most of the population

* Camillo Benso Cavour (1810–1861). Involved in the 1830 revolution in France. Later Secretary of Economy in Italy. A liberal economic reformer.

was concerned, it was also the poorest of the large cities. All considerations concerning an economic basis for a scientific institute in that environment had to be corrected by simple statistics on the problems of income, degree of and desire for education. The few people who knew something about Italy, whose advice or support Dohrn could obtain, made it clear to him that Naples was »a particularly difficult« location.

When he left for the south toward the end of February, 1870, (his parents did not know of it) he had nothing with him except a few notes concerning lots for sale; the Baranowskis had received a few offers from acquaintances. He first went to Florence where, during the past year, he had made friends with a Hungarian refugee, Gustav Frigyési. When the latter was 15 years old he had taken part in the Kossuth revolt of 1849 as a drummer. Then he deserted the Austrian Army during the 1859 campaign and advanced in Garibaldi's army. He had distinguished himself and belonged to the close circle of the leader of the insurgents. Dohrn had made contact with him through the Stahrs. Different though their origins and fate had been, mutual understanding was easy. Frigyési had told Dohrn that he would help him with recommendations. But when Dohrn arrived in Florence, he found the man in the midst of total financial bankruptcy. »Frigyési too had joined the ›entrepreneurs‹ «, as Dohrn wrote. He had founded a Society for Colonization in Sardinia and was now, with many debts, at an end. Dohrn should help him. It is partly moving, partly funny that Dohrn's first act in Italy was to address German capitalists – in vain, of course – to help the Sardinian colonization. The humanitarian idealism of the follower of Garibaldi coincided with Dohrn's easily inflammable enthusiasm. But first, he gave the man 50 Thaler from his own travel chest so that he could eat and flee to Switzerland in order to escape prison.

Frigyési was the first character in an adventurous line-up which was part of the early history of the Zoological Station – others followed. But he provided Dohrn with several connections with people in Naples and, what was most important for him, Garibaldi himself wrote a recommendation to his Neapolitan confidant. With that he might hope to progress. But this document was confiscated – Garibaldi's entire correspondence was watched at that time. Dohrn only discovered later to whom it was directed.

Above all, the trip to Naples was supposed to clarify the problem of the lot, if possible even to settle it. It soon became clear that the various houses offered were either too far removed from the center of town or too high above sea level. And of course, nobody knew what an aquarium was – London, Hamburg, Berlin were the only places which owned one. It seemed wise to knock at the door of the university zoologist. That might have ended badly if he had been a man afraid of competition with his own work. This was, however, not the case with Professor Panceri. * The young German was amazed that the Italian scientist did not have any idea about the revolution of biological problems in England and Germany. He was a self-made man who had relied, in his education, on French textbooks; their prevailing direction remained for some time strongly opposed to the »amateur« Darwin. Although Dohrn's inmost scientific desire remained, in his own words, »almost incomprehensible«, the man was nevertheless won over to the plan. His dignified loyalty helped; it grew to become a reliable friendship. Later, he graciously accepted that he, the representative of biology in Naples, remained in the background behind the intruder.

This first visit happened by chance at the best moment: Naples was planning a maritime exhibit. Panceri had been asked for advice, but he was somewhat helpless and took the German as his substitute to the planning committee. The site planned at the beach of Mergellina** was favorable. Dohrn boldly aimed at his whole plan. He suggested to the Commander del Giudice that he would

---

* Paolo Panceri was professor of Comparative Anatomy, but *not* Zoology, at Naples University.

** A Western suburb and harbor of Naples, at the foot of Posilippo, a promontory West of Naples.

like to build, at his own cost, an aquarium for the period of the exhibit, the income should go to the Committee – that was meant as a special bait. Afterwards they should leave the place to him – the apparatus of this improvised enterprise would then be at his disposal for the planned station without new expense.

The proposal to gain something attractive without cost impressed the director of fire protection – that was del Giudice's position. But the question of a permanent place was a matter for the Sindaco.* He was willing to introduce Dohrn to him. But before this happened, the whole idea went to pieces: one day, Dohrn received the bill for sheets of glass in his house that he had not ordered, measures he had not calculated. Confused and furious he went to del Giudice, who had gotten the Committee together, and made a scene which, not as tactful as the whole engagement had been planned to be, broke up this bridge to success and created his first enemy. He had to assume that some profiteering was going on. The few Germans who lived in Naples had urgently warned him against the whole enterprise (the doctor Schrön, the Consul General Stolte) because he would lose his money and temper. Perhaps the attempt might succeed, if he could find some influential men who would also make some money. But Dohrn, following his own nature, neither could nor would take this route. Panceri considered the conflict irreparable, which had ended, very dramatically, with the slamming of doors and walking out. But the stalemate was suddenly broken, when Dohrn produced a little sketch. With a few strokes of the pen he sketched the building as he saw it in his mind; pleasant, simple architecture, with good measurements and loggias producing shade; the organization of the building had accompanied him already in Jena. Now here, when he saw the spatial possibilities, he developed a façade. Without having been professionally trained in sketching, he had a clear and sure stroke, thanks to his scientific studies; and his inborn feeling for artistic order gave him, as he tried, a wonderful sureness. Panceri

* Sindaco is the mayor of a city.

held the piece of paper in his hand for a long time, then he suddenly said: »Your plan is so handsome – this building should stand in the center of the Villa Reale.«* And he offered to go together with Dohrn to the Mayor he knew, the Sindaco Count Capitelli, and put the matter before him.

Later, one of the often told anecdotes was that the architectural sketch opened the door to the Municipio. ** For the official presentation, Dohrn had redrawn it – and the result was good. The scientific impulse of the enterprise did not interest the Count, but he, too, liked the sketch. And they could tell him in good conscience that the future Naples Aquarium would surpass all existing edifices in richness and splendor. He charged his assessor, Marquis Cedronio, with the further negotiations concerning a fitting building site. It was an enormous gain that the Villa Reale, the beautiful park at the beach, became the center of discussions, without Dohrn's giving the first impulse. The general traffic problem was favorable, the sea reached up to it, in fact, it washed against it. There was, however, one thing Dohrn did not know: the enterprising city administration had already contracted with a Belgian conglomerate to push the sea back, widen the beach, and build a large new road along the coast up to the Posilippo. Whoever sees the grounds now must reconstruct in his imagination the former situation, when the sea reached quite close to the Station. The fight about this project had as its background a large speculation about the grounds and the opening up of the coast between Mergellina and Posilippo. None of this was favorable for the construction of the project in the coming years. Dohrn could not foresee this, just as, luckily, he coud not foresee many other things.

The optimism following the negotiations at the Municipio seemed justified. The skepticism of the German Consul General changed to sympathy after he received a personal recommendation

* Villa Reale is the large public park in which the Station was built. Today: *Villa Comunale*.

** Municipio is the city government.

from the Ambassador of the North German Union, Count Brassier de St. Simon. Dohrn had requested it through the well-meaning collaborator of Bismarck, Lothar Bucher. At the same time, Frigyési came up with various introductions (received from Signor Giuseppe Dassi) to some scientists and to the head of the opposition, the impoverished Duca di San Donato. For Dohrn it was the second lesson in finding his way through the Neapolitan machinery and of collecting anecdotes. Only the surgeon Palasciano understood what it was all about. Later Dohrn no longer needed to negotiate with him, but the man remained »constantly a moral support«. Now he needed him, particularly since an enormous, active, loud helper appeared, a broad-shouldered giant with a long beard. When Giuseppe Dassi, sent by Frigyési, stepped on the stage of action, it seemed for a while that he was the main figure of the drama, whose plot was not yet very clear. His visiting card called him »Ingegnere, Milano« [Milanese engineer]. It also noted that its owner had been, during Garibaldi's Neapolitan period, a deserving civil servant. Dohrn never succeeded in understanding this man very clearly, who suddenly appeared »like a meteor«. The stormy, matter-of-fact behavior with which Dassi usurped every confidence without being asked, opened doors even if the people inside did not know him; how he decided to publish a journal immediately, called *L'Aquario,* for which he had already found a translator for Dohrn's essays – all this intrigued Dohrn's psychological curiosity, but it also trained him in being cautions, since by now he had already had some experience. It became a kind of hide and seek game: was he an adventurer, a businessman, or just a character who amused himself with noise and eagerness? But Dassi had written letters to the Italian Department of Education in Florence. Assurances arrived from Minister Correnti and his leading civil servants that might become useful, and the matter had thus been officially raised above the merely local level.

In between Dohrn had visited his friends the Baranowskis in Messina, a trip which made him feel good. There they believed in him – at that moment this was important, because the reaction of his

German friends was not optimistic. The Stahrs had not refused his request for engaging their manifold connections in the South. But since they had already experienced the exuberance of Dohrn's self-confidence, they were more scared than happy with his reports. They also worried constantly about the relationship between Anton and his father. Fanny Lewald reproached him for his »fantasies«. She demanded something »positive« and called it »unhealthy« that he, who was still completely entangled in his Neapolitan problems, was already talking about stations in Venice and Nice. When Dohrn read these letters again later, he must have understood them. At that time, however, he was much hurt. »You accuse me of fantasizing, and I accept the word. For while it is at times mistaken, it is also an advantage. Without these fantasies we would still be at the same point . . .« He simply rejected the »reproach of daydreaming« and would not ask them for any further help, »not even a recommendation . . . If we do not understood each other in this matter, the fault may be found in our long separation and the great gap between our philosophies. Where you feel horror, I feel happiness and joy; where you don't want to think or look, I prefer to live and like best to think within the vast dimensions of my thoughts, whereas you start from the minute situations of life. Let me go, rebuke me less; I will react more readily to the pressure of a silk thread than of a horsewhip, if there is ever need for correction . . . all is going well and I owe my thanks to *myself*. That makes me happy and will satisfy you. You will find me independent in ways which may be unexpected to you« (May 5). The Stahrs had often stayed in Italy and were considered particularly knowledgeable about the land and the people (hence their worry whether Dohrn would find his way). A little later they received this wise counsel: »Believe me, the fact that you traveled back and forth through Italy for ten years does not let you see more deeply than my quarter of a year of real and serious business.« Most people fail in such an attempt. As he had already said in his long letter, he owes his progress to the ability »to speak to all people in their language rather than in my own«.

Thomas Huxley had answered, with great warmth, the report that a series of stations was to be created. He recognized, of course, the basic importance of the plan for the development of science. He sent Dohrn an invitation to the British Meeting of Scientists of that year and promised the warmest support. This was the natural occasion to take up contact with his father again – at some time and in some way the relationship had to be put in order. A letter Anton had written from Jena still remained unanswered. Of course, Huxley was not Alexander von Humboldt and he had not been asked for help in personal matters. Nothing had been farther from Dohrn's mind – but Carl August Dohrn knew the scientific and public rank of this man. His own friendly connections with British scientists were so lively that he knew how to value their interest. In a long description which presented what had been achieved and what was still planned, Anton explained the matter to his father. He added Huxley's letter and explained that the next stations he had in mind were Venice and Nice. This was a mistake. Full of imagination as Carl August was, speculation and the founding of new things were not to his liking and these he suspected in the »daydreamings« of his son. The result of this »reconciliation« remained doubtful. His first judgment, addressed to Adelheid Dohrn, was rather laconic: »Read this, the boy has become insane.« This letter, too, the father, who liked writing so much, did not answer, and the mother stepped in. She thought she felt that, even if the old man took shelter against the revolt of the son, he was still impressed by Anton's recklessness and the recognition by others. It might be possible to predispose the father more favorably. (Anton wrote to Stahr that he thought the »reconciliation« had taken place.) Adelheid Dohrn had for a long time had free disposal over the income of her husband, who had very little interest in business. She saw to it that her son was not quite without financial help during his negotiations in Naples. But when the father finally answered the request to pass on to Anton, as he had done to his brothers, a capital of 30000 Thaler, he stuck to his »No«.

This happened in a situation which might easily have broken the son. Anton, together with the Marquese Cedronio, was looking for a building lot. Intermediate plans to take over existing buildings in Villa Reale came to nothing, since their present rental agreements were too valuable to the city. In the meantime, Dohrn's inner image of the projected Station gained more and more shape for him. He had talked with Alfred Brehm, the founder of the Berlin aquarium, about some of the technical needs. This he developed further in Naples. The letters to Adolf Stahr give a good insight into this. The plan included not only an apartment for himself and of course a laboratory and a library, but also lodging for ten or more strangers. The number of separate workrooms, namely four, seems small in comparison. The large laboratory on the floor above the public aquarium was supposed to be the most important room. The size of the whole building depended on these rooms, as calculated according to necessity. When this was later carried out, these plans had to be dropped almost completely. But it proved to be fortunate that they were conceived somewhat grandly when a local architect, Signor C[apocci], drew the plans for the official building request. The somewhat worried Dassi had found this man among his clients, and he later proved to be a difficult case. As far as financing the whole thing was concerned, some of the plans had changed: the matter would amount to approximately 36 000 Thaler. Dohrn counted on only 20 000 to 25 000 Thaler from his father; the rest he hoped to raise among friends. The idea of shares was given up: »I do not want to use shares, in the first place because I would like to determine dividends myself, and secondly because I don't want anybody else to have a voice.« Carl Vogt, who had more experience in business practices, had told him in detail in a letter that capitalists give away their money for dividends rather than for scientifically doubtful enterprises. A handsome illustrated catalog with pictures and with popular scientific explanations appeared in the calculations of the current income. Then, in a letter to Fanny Lewald of May 21, 1870, an idea occurred which later became very important: »Furthermore, we have to count on the sale of live animals to aquariums in Berlin, etc., and of animals preserved in alcohol – something, I estimate, that will bring 4000 to 5000 Thaler.« This sounds as if Dohrn had gradually gone over completely to the field of commercial calculations. But the same letter is still concerned with the scientific situation and tasks. Kovalevsky worked on the same embryological matters. Independently of each other they had arrived at similar results. This satisfied him very much »with regard to the Jena colleagues«. Since Darwin's theory was, after all, English property, »it was important to secure for the Germans at least the discovery of the entire animal genealogy rather than leave it to the Russians.«

The search for an appropriate space within the Villa Reale found a satisfactory solution in connection with the planned pier edifices. But the manner in which the property was to be transferred caused fundamental consideration. Could or should the city give up its right of ownership to this spot to a private foreigner who had, aside from official German and some more imaginary recommendations, nothing to show except the assurance of his behavior? For what he really wanted, the stranger could not expect understanding. He understood the official hesitations. Later Dohrn sometimes used the comparison: how would a young Italian have fared who appeared in Berlin with the wish that one might give him part of the Tiergarten?* When he sensed the hesitation, he had a surprising suggestion: Let me have the site now, without cost, and after my death (or in case I die unexpectedly, after 30 years), at the end of the century the building with its contents would fall back to the City. The Sindaco Count Capitelli was not disinclined. Dohrn figured that in this way he would not have to pay rental debts for the ground. When one started to consider the matter juridically, the question necessarily emerged: What is the position of the Municipio – if Dohrn should die earlier – with regard to his heirs? He had no heirs of his own and had, in the program of his life that he sometimes formulated, not foreseen marriage. Should the parents

* The large park in the center of Berlin.

or siblings in the far north become partners of an Italian city administration? For them, it would be as doubtful as for the Municipio. Dohrn, who did not intend to invest in a family possession but rather to give it to »science,« found the solution in appointing as his heir »a German university,« and, when the partner added »or an Italian university« (thinking of Naples), he was diplomatic enough to agree. The tone of the contract was rather idealistic, and this elementary quality remained its permanent characteristic.

It also contained problems not realized at that time. Still, Dohrn had learned from the experiences of the previous months to have enough confidence in himself to be able to navigate his boat through the rough waters.

The final formulation of the contract brought unexpected and paralyzing difficulties. Capitelli had instructed the Secretary of the Municipio, Cammarota, to take care of it. This man later became, throug his practical loyalty and human friendship, the most important support for Dohrn's activities in Naples. Now a paragraph had come from his pen which forbade anybody to live in the future building. Why? That was the regulation for the Villa Reale. Furthermore, the building was forbidden to have a kitchen. Why? The enterprise should not gradually turn into a hotel. Finally, »women were strictly forbidden to stay there.« Why? The house was also fit for a bordello – one wanted to protect oneself from this. This kind of thinking was not expected by Dohrn, who had already objected to some other regulations. Red-faced, he screamed at the Count that anybody with self-respect could not accept such suspicions. The contract with such articles would not be signed. The Count said: »Then there will be no contract« and Dohrn left, angry and amazed and almost somewhat amused that the efforts of these weeks should be entangled in such a mad grotesque.

Signor Dassi was also furious. This would have been impossible if the followers of Garibaldi had ruled the city. But he would ask the Department of Education to intervene – which he did. Dohrn never forgot the confidential answer from Florence: »Please tell Mr. Dohrn that he should sign everything, and afterwards do whatever he likes.« The advice of the very pragmatic ministerial administrator probably indicated his knowledge of the place and the period, but he knew nothing about the young German. The latter did not want to put up with the idea that he should reach his goal under the distrustful suspicion of a disgraceful speculation – the well-drawn-up contract remained unfinished. In early June Dohrn traveled north toward home by way of Nice and Geneva, where Carl Vogt encouraged him. At home, the difficult task awaited him of clarifying the financial situation. At first, the reports to the Stahrs from Stettin seemed satisfying: »My father tried first to play the offended party; but since I absolutely disregarded that and instead very gaily told all kinds of stories, he too became serene and talkative.« The son believed firmly that he could count on the necessary money. But then a »lively animosity« returned, so that Anton Dohrn went to his friends in Jena.

He had, of course, asked for leave again during the summer semester, and his problematic situation as a lecturer there oppressed him. Now, it became more a question of resignation. His father's wish that he get a professorship before making any further decisions had been by-passed through his independent negotiations in Naples. His situation at the faculty of Jena had not improved. He knew that the »evil word« circulated among his colleagues: »Dohrn is great in all breadless arts« – that referred also to his social acts – »in the bowling alley, in chess, whist, and L'Hombre, etc.« (as he himself noted with self-irony). More importantly, however, the scientific confrontation had deepened. As he formulated it, »the predominant conception, represented mainly by Gegenbaur and Haeckel, saw in the Ascidiae and the Amphioxus the lowest vertebrates; from them evolved in an unspecified way first the fish and then the higher class of vertebrates together with mammals and man. In contrast, I had gained another insight, namely to derive this higher development from the annelids, the ringworms. This view put me in the strongest opposition to my two former teachers. And as things go, this purely theoretical contrast led to a personal opposition which was nourished by a difference – existing also in other ways – between our respective natures . . .« Jena had become »uncomfortable« for him. And the justification of his own point of view, he realized, needed very long and special work, for which Jena lacked the preconditions. In Naples he could hope to find it. That was the hidden feeling which drove him to the southern adventure.

With a distrustful, perhaps even malicious curiosity, one might have watched in Jena how the Privatdozent Dohrn began to tear down his position before it was even very secure. This was not only true for Jena. He came to feel deeply the disapproval of the enterprise, the critique of his lack of business experience, particularly by people who were close to him. He was angry about the fact that they considered him a Don Quixote. Later, after he had gone through crises and tribulations he could or would not as yet foresee, he understood this blaming or cautioning, worried or mocking attitude. Yet, after three decades, he wrote in a typical self-analytical retrospective account: »Two qualities remained with me which protected me from a catastrophe: an instinctive art of delay and a considerable versatility in finding ways out and detours around problems toward better situations. Undoubtedly I had, and still have, the ability to jump with an apparently quite one-sided delusion to a conclusion which may seem very doubtful to someone else. My lively imagination and my need for communication, connected with this, may easily give the impression that I look neither right nor left, but, as if hypnotized, only in one direction. In fact, this is not the case, although I appear thoughtless and speak inconsiderately. Nevertheless, it is easy for me to turn around completely after having satisfied my need to talk and have followed the one-sided conception to the utmost. Then I can follow up a diametrically opposed line of thought, and if possible reject my first conception even more thoroughly than my original critics. Almost always a certain one-sided fanaticism in grasping all possible points of view was followed by a careful consideration of the results which I mainly aspired to and expected. More than that, I was instinctively on my guard against the side effects, disregarded by others and apparently unimportant, which only too often become the source of essential disturbances, even if the main goal is achieved.«

3 INTERRUPTION BY THE WAR

The Franco-German War which started in July 1870 overthrew all plans. »Don't worry, Papa Stahr, I will pay my blood tax. Several days ago I signed up for service in the Army. My heart bleeds when I put on the thin blade of a sword all the good and noble things that I wanted and was able to do – but what can I do?« (July 23, 1870)

Because of post-operative complications of an appendectomy, Dohrn had been placed in the reserve, at the second summons. He had expected to be dismissed from military service after 1868, but when he reported to the military in Weimar* he was told that this was not the case. Since he had already missed two control assemblies he was told to go immediately to Kassel to the training battalion. There his short military service took place. One of his memories was that he stood guard there (together with others) over the imprisoned French emperor. His attempt to join a medical troop did not materialize. Since he was a »learned man« he was assigned to an office.

The lively political reflections from the years of the Prussian conflict, the campaign of 1866, had diminished in his correspondence. In the exchange of letters with F. A. Lange the latter's amicable effort to found a committee of »The United States of Europe« was being discussed, but the warrior-like passion was somewhat weakened as his image of the world became broader. Dohrn had appreciated, during his second visit to England, the respect the Prussians enjoyed and the English dislike of France (not only of Napoleon); he could report similar experiences from Italy.

* Weimar is near Jena.

But it is characteristic of the structure of his thinking that he tried to understand the Franco-German hostility in its European context. He wanted to clarify this situation for himself, mainly in his correspondence with his friend Katharina von Baranowska. He was not sure where the Polish patriot's sympathies lay, and whether she expected reactions and relief for her own people (which he did not consider very likely). But »I wish it out of a sense of fairness and of my warm friendship for you.« On August 8, after the days of Wörth and Spichern,* he wrote to his friend: »A new era is beginning for Europe. The nations will be freed. The despot and the Pope are defeated, the ground belongs to us, and there is so much intelligent and revolutionary spirit in Germany that in a short time nothing of the Middle Ages will be left in Europe.«

He agreed with the fight against the hated Napoleon – after the imprisonment of this emperor the war seemed to have lost its meaning. Dohrn believed that it would soon be possible to come to a decent peace with the newly founded republic. But the nationalistic tendencies of the Germans, by which he felt surrounded, worried him. »People who think as we do are few and nobody listens to us.« And sorrowfully he wrote to Huxley, with whom he had just talked: »One could really be afraid that the good people esteem killing higher than working. I will be happy if I can escape the next years and go to Italy, in order to uphold German nature and inclinations better from there than in the midst of the politicians, who have become intolerant and anxious to be successful.«

The meeting with Huxley just at that moment became very important. The English scientist had invited Dohrn to participate in that year's meeting of the British Scientists. One could hardly think of this after the outbreak of the war, and Huxley himself did not count on it. »If you Germans do not give that crowned swindler** (...) such a blow that he will never recover from it«, he had written on July 18, »I will never forgive you«. But when Dohrn

* Two early battles in the Franco-Prussian war of 1870–1871.

** Huxley means Napoleon III.

realized that his various efforts to find another military position would not soon be successful, he asked the minister and commander for a leave of two weeks outside of Germany. This was granted, and he traveled to Liverpool as soon as possible. Huxley was so amazed and glad when his young German friend entered his room that he completely forgot to open the meeting, which was waiting for him. The German soldier on leave, who had several acquaintances and friends in this circle from his visit at Dundee three years ago, had the effect of a sensation. The Cambridge professor, Foster, gave him a public reception as a »ghostlike apparition,« and they applauded when he came in. He felt he was in a friendly atmosphere, and reported to Stahr that »from Darwin down to the last Englishman I saw, everyone spoke, thought, and acted with one voice and complete devotion to Germany.«

The main thing, however, was that he could speak of his plans in innumerable conversations – they were »enthusiastically received«. The general assembly unanimously accepted a resolution suggested by him: »The assembly declares that the foundation of zoological stations in different parts of the world will be of the most significant influence on the progress of biological science, and it considers the foundation of such a station in Naples as the first decisive step in this direction.«

Similar declarations were promised him by the Royal Societies in London and Edinburgh, and by the representative of the Belgian Academy. This military leave gave him more than he could reasonably have expected. The fact that such a well-known English corporation stood behind him and called together a special committee for this matter, whose regular member he became, had to be helpful even at home.

These 2 weeks were thoroughly productive. He went to Robertson, the old practitioner in Millport, in order to be confirmed and advised. He dealt with glassblowers about quality and price for the windowpanes needed for the aquarium. This time he did not object when Huxley introduced him to Darwin. [20] Darwin knew about his enterprise – almost a year earlier he had answered Dohrn, who had written to him about this, in a friendly way, but with the advice to be cautious. In his records Dohrn writes extensively of this visit, »which left a long-lasting impression« (September 26, 1870): »At 11:30 Darwin entered the room in the company of another gentleman whom he accompanied to the door; he took my hand and apologized that he had kept me waiting, but he returned soon after having said goodbye to the other gentleman and said, smiling: ›You will think me very respectable; this gentleman was our clergyman!‹ – indicating that he was not quite an outcast and had friendly relations with individual clergymen. Then we went into his private room. I admit that the personal appearance of Darwin surprised me very much. I had expected to find a sickly-looking man, but I found before me a tall, strong, gray-bearded person, full of life, serenity, and a heart-rending friendliness. The conversation began, understandably, with the enormous historical events which had just taken place. Darwin showed the greatest sympathy with the turn of events which had dethroned Napoleon III and had severely punished the French arrogance and carelessness.

»The conversation took a turn which, as an anecdote, is too characteristic to leave out of this story. At that time the Prussian Uhlans * had achieved some outstanding success, so that they had reached a kind of legendary importance. Darwin asked, ›Tell me, Dr. Dohrn, what are the Uhlans?‹ I answered, jokingly, ›An Uhlan is standing before you, I have been serving for six weeks in an Uhlan-regiment.‹ Darwin was highly perplexed by this communication, and when I asked him what he had believed the Uhlans to be he answered: ›I thought, they belonged to a half civilized tribe on the eastern frontiers of Germany.‹ I smiled, of course, and answered: ›Your view is no compliment to the eastern neighbours of Germany!‹ Even later, in a letter addressed to me (which is reprinted in Darwin's biography) Darwin wrote: ›I have often boasted that I have had a living Uhlan in my house!‹ Darwin took the liveliest interest in my Neapolitan plans. He was happily look-

* A Prussian regiment.

ing forward to their eventual accomplishment, and he then talked to me about his own scientific work. In the course of this conversation I asked, ›Excuse me if I ask a perhaps very stupid question: how do you begin your studies at all?‹ ›I'll tell you,‹ he said. ›I begin always with a priori solutions, if anything happens to impress me. I have generally hundreds of hypotheses before I know the facts; I apply one after the other, till I find the one which covers the whole ground: but I am exceedingly careful and slow in printing!‹ We talked for a full hour and a half, and then Mrs. Darwin entered and took him away because he was never supposed to talk any longer than that. After half an hour's rest he came back and we went to have dinner. Then I said goodbye, not without having to promise that I would return as soon as possible.«

When Dohrn had returned to Kassel, he heard that he was for the time being discharged together with a group of older men. (»Because they have absolutely no use for me,« he explained in a letter to Grant.) The situation was uncertain. His letters show that several times in the course of the winter he expected to be recalled. After the strains and the excitement of the past year, he finally fell victim to a deep depression which made him unable to promote his plans at home. He remained in Jena, carrying out research on pike and perch eggs, but without much enthusiasm. He wrote to Grant: »Now let us go to Naples; let us hope that the world behind my skull soon finds its balance again, and also that the outer world regains its balance as well – then we will found a common and decent existence for ourselves.«

4 UNCERTAIN PRELUDES

In the days of July 1870, shortly before Anton Dohrn had to report to the Army in Kassel, he received a letter of credit from his father for 12 000 Thaler. Through the good offices of the Consul General Stolte in Naples, it was placed at his diposal. Carl August had added the following letter: »I would lie if I were to say that, after these oblique and crooked attempts, I could look at your project as being in any way certain. But you have started it, and I hope that your dream does not bring you more hard knocks than successes. As long as the war trumpet is blown one cannot foresee its changing into a peace pipe – not if France persists in its obvious insolence. Therefore, the prognosis for the whole enterprise is not good at this moment. But you are yourself responsible for the fact that I did not, as I would otherwise have done, oppose it completely. Perhaps you will listen to the advice I am giving you for your own benefit and will restrict your plan to one place for the time being; sanguine, indefinite speculation without a powerful sum of money behind it carries ruin with it and would force me, on account of your siblings, not to pass beyond a definite sum.«

This produced a certain clarity. The sum was not half of what his son had expected, based on what his older brothers had received – an amount on which he had counted to accomplish his enlarged plans in Naples – but for the moment he had to accept matters as they were. He knew that, after all, the sum did not constitute too much of a »sacrifice« in the context of the estate which grandfather Heinrich Dohrn the elder had founded. He prepared himself mentally to continue this kind of exchange, although it was depressing. The plan itself had become so powerful that his moral self-assurance depended on the venture.

The plan itself would soon have to endure new burdens and dissappointments. An earlier experience was not particularly encouraging: a very wealthy, childless relative whom he wanted to win over to his Messina plan had sent him only 25 Thaler! But in the early summer of 1871 he still hoped to raise the building fund from his friends – the result was completely negative. Why did he not want to sell shares? That had become very fashionable in Germany at that moment – it was not more speculative than many other plans to use entrance fees to an aquarium or zoological garden for capital gains. From a business point of view this was worthwhile advice, but it missed the saving thought conceived on the carriage seat of Apolda: the entrance fees were not meant to produce dividends, but should serve as support of the scientific laboratories. The donation would have to be à fonds perdu – was the young German nation financially mature enough to give money for abstract science? He was told that this was a matter for the State. He should go to Rome and tell them that Prussia had promised 25 000 francs if Italy would donate the same amount; in Berlin he should reverse the proposition – then all would turn out well. That was a clever thought. Later, Dohrn may well have remembered this bold suggestion when, not quite in this crass fashion, he managed to play governments against each other for this cause, with regard to good will or aloofness.

In the beginning such an idea must have seemed to him like mockery. Would he find greater willingness in England, where he had been given, in Liverpool, such friendly (and certainly valuable) support for his plan? He warmed to the legend of the private Anglo-Saxon willingness to have an open purse for things of common use and of science. Thus he went there to talk the matter over with Huxley. The latter was as kind as ever, and the freshness of the country made him feel good enough to rid his mind of the shadows. But his friend made it clear to him, with great amicability, that he should have no illusions that he might in this country meet the rich enthusiast who would donate money to a German so that he could live in Italy for his own scientific preferences. Thus the plan appeared to be impractical for the moment.

It was easier to find moral support than financial. But it was of course not useless to try. Before Dohrn prepared for his new trip to Italy he asked some leading scientists for expert opinions: H. Helmholtz, E. du Bois-Reymond, the Leipzig zoologist Leuckart – all were ready with warm recommendations, Leuckart even with enthusiasm. His old friend Carl Vogt described his own efforts, and even Gegenbaur and Ernst Haeckel were willing, the latter with extensive suggestions. Dohrn wrote a short, restrained introduction to these letters which he enlarged into half a Bogen * – the first

* Bogen is the unit of measurement of paper for printing. For that purpose a Bogen is folded into various sizes (hence the terms quarto, duodecimo, etc.). A half Bogen may therefore have varying numbers of pages.

publication for this plan. The weight of the names balanced the fact that in Germany no official society appeared as it did in England. The position of these men lowered even the distrust in Hökendorf. Carl August sent the little booklet to Karl Ernst von Baer in Dorpat whom he revered and in whom he won for his son a warm advocate, even in dealing with his father.

Since Anton Dohrn had left Naples in the early summer of 1870, matters in Naples had become nebulous. In December 1870 he had effortlessly managed to have the Prussian Minister of Culture, von Mühler, advise the General Consul, Stolte, to monitor local progress and to take Dohrn's part if necessary. Stolte, however, could only report on the chaos in local politics, something his skeptical attitude had made apparent to Dohrn earlier. And he warned him of the »efficiency« of Signor Dassi. When this somewhat mysterious follower of Garibaldi pressed him somewhat hard with political letters, expecting answers from Bismarck (as if Dohrn were a political agent) Dohrn felt more and more uncomfortable. All this could be clarified and ordered only personally.

In the first place, he wanted to prepare the technical matters. As things stood in Naples, the technical development could not be taken care of. It was a modest satisfaction that Signor del Giudice's attempt to build an aquarium had become bankrupt both financially and in its construction. Dohrn had completely broken with him.

For some time work on the construction of aquariums was mainly done in England. Their first plan was based on the theoretical understanding that by a correct balance between animal and plant life in a closed vessel, water would always remain fresh. Practically, however, this theory of balance was inadequate. It was impossible to calculate the production of oxygen of plants and the excretion of carbonic acid by animals in such a way that they corresponded to their mutual needs. The lighting needed for the public to watch brought about too quick a development for the microscopic germs of algae, and this destroyed the assumed balance and also clouded the water. W. A. Lloyd of Wales completely changed the theoretical system by a few practical steps. He did not care about artificial designs of planting. The growth of plants in water was regulated by limiting the entrance of light. In this context he found that the solution was to keep the public rooms in darkness, which naturally increased the effect of colors in the marine fauna. But this was not the decisive theory; rather it was the artificially created addition of atmospheric air: out of a dark reservoir, water was added to the basins and air was mechanically mixed with the streaming water under a certain pressure. The stream which was achieved by way of circulation took the dissolved air bubbles into all corners of the basin, took care of breathing, and simultaneously of the oxidation of animal secretion.

He was the first to build the Hamburg Aquarium according to this principle. It was basically a success, also as far as financing the enterprise went. He was still in Hamburg when Dohrn was there in the summer of 1867 studying the eggs of the Norwegian crab; he was not happy when he was told by the Director of the Zoological Garden to lend a hand to the young man. He was a practical man and not at all interested in theoretical embryology; he considered the whole thing as a disturbance of the daily routine. That changed one day, however. Dohrn reported how, early one morning, he went down the aquarium steps whistling. Suddenly Lloyd put his head out of the door: »Do you know what you are whistling?« he asked.

»Of course I know.«

»It is the violin concerto by Mendelssohn that I like very much.«

»So do I, and Mendelssohn is also my godfather.«

Dohrn later reported that Lloyd was close to embracing him, and said over and over again, »You are Mendelssohn's godson! Then you are welcome, we will be close friends.« This intermezzo was the beginning of a close relationship and, as we shall see, of an unselfish friendship which became very important for Dohrn.

At the time when Lloyd was working in Hamburg, Alfred Brehm was Director of the Zoological Garden. The two men did not get along very well. Brehm went to Berlin in 1867; the reader will remember that for a while Dohrn had been considered for

Brehm's successor. Brehm founded the Berlin Aquarium and, with a few technical changes, followed the principle Lloyd had developed. The latter was called back to his homeland in order to start an aquarium in London. Dohrn had talked with Brehm about his plan as early as January, 1870; he took up the relationship in 1871. That was a good plan, because Brehm had behind him the experiences of a new building and he did not hold back his best advice, as this man who stood before him obviously did not have sufficiently clear ideas of the difficulties awaiting him. He suggested to Dohrn to secure the young Swiss architect Kelterborn, who had directed the construction of the buildings in Berlin. That made sense. It was possible because the Swiss had time and interest. He also knew how to avoid mistakes in organizing the delivery of supplies, etc. This was the first man in the staff Dohrn was to gather around him. That this would develop into a half-comic, half-tragic novella he could not foresee. The company of a German expert appeared to him as a good support in the negotiations about the building which might arise with Signor Capocci.

5 THE BEGINNING IN NAPLES

Dohrn returned to Italy in October, 1871. He arranged his return in such a way that he could participate in the second international Congress of Prehistorians in Bologna – this would offer him an opportunity to become personally acquainted with some Italian scholars. Germany had sent a single representative, »only one, but a lion«, Rudolf Virchow. Dohrn had once been his student and it so happened that now he became something like his assistant. He was present when there was a conflict between the French delegation and the German scholar, for whom the Italians then prepared a triumphant rehabilitation. Virchow's knowledge of Dohrn's plans was later helpful for their scientific advancement. It also happened later that Dohrn had to complain amicably about the man, who was very accurate and rigorous in his parliamentary activities. Now, in the company of Virchow, Dohrn had no trouble finding the hoped-for connections. The most important of these was the relationship with the Bologna zoologist, Trinchese. Later, he became Panceri's successor in Naples and, like Panceri, did not treat the intruder as a rival but rather supported him as a friend.

When Dohrn arrived in Naples, good and bad things awaited him. Among the good things he counted the presence of the Baranowski family. Mr. von Baranowski had given up his position in Messina and had returned to Moscow for financial and industrial activity. The delicate situation of his wife's health recommended a further stay in the South. Their attempts to stay in Vienna or Warsaw indicated this and the decision favored Naples. In addition to the better medical care, the wish of the German friend was not without influence. Katharina von Baranowska heard the tone of homage in his letters: »The relationship with you and your family has the value of a poem for me, one which I read and re-read every day, and which I quote when I cannot read it.« The receiver of the letters, a lively and sensitive woman, only a few years older than this interesting, educated young German with his somewhat romantic idealism, was not unreceptive to such words as she received in the winter of 1871. With her two daughters, she had taken an apartment in the old Torlonia palace at the Mergellina. On the upper floor there were still a few empty rooms available which Dohrn rented for himself and the architect. He also took the young English zoologist, Ray Lankester, into the apartment. Lankester, who later became a professor at Oxford, had planned to work that winter in Naples anyhow.

They had the old boxes of the portable aquarium which were still in Messina sent to Naples. Dohrn was looking forward to continuing his own studies together with the Englishman, who was a cheerful person. But things turned out differently.

There was a sequence of adventures, uncertainties, vexations, catastrophes – the »hot-blooded recruit of life and science« had no time left for the microscope. He was changed into a »perpetuum mobile« and ended up in a chaos of worries and responsibilities.

How, when, and where would the building be started? Kelterborn made sketches and calculations, but this remained paperwork. In the meantime the architect fell in love with the 16-year-old Marie von Baranowska. An excited letter to Dohrn, asking him to woo the girl for him, made the situation grotesque. Friendly kindness helped to overcome the painfulness of the situation, but the young Swiss had a nervous breakdown and some countrymen had to take him to a Swiss hospital. Although he was cured in a few months, circumstances made it impossible for him to return to his task. The most important support for the building technique had broken down. The new Italian architect was less eager, but he proposed to Dohrn something wonderful for presentation to the Municipio. Finally, the large design was finished. He had it carried to Dohrn and while he himself loudly appreciated the appearance of his work, Dohrn (who after all had ordered it) sat in front of it embarrassed and whistled to himself, this time not Mendelssohn but Beethoven. He should not have done it, for the artist took his hat and said goodbye. Two days later Dohrn received the visit of an Italian gentleman, who brought him the challenge to a duel with the architect, who felt that his honor was deeply wounded. Dohrn was by now so used to the play of paradox that he did not lose his sense of humor. There were a few dramatic scenes, between calm serenity and fiery emphasis – and then the hurt honor was restored. Now a real contract was made, with dates for the delivery of the plans, with fees, and an emergency amount in case something went wrong. Dohrn reserved for himself the choice of the builder.

That was a very intelligent move. He had burned his fingers often enough, and more caution was indicated unless he wanted to risk losing his way among the hostile cliques which he was told were blossoming all over Naples, particularly in this field. Through the General Commander of the Army he heard the name of Giacomo Profumo, who had once been a captain in the Engineers. He was more expensive, but he was also more reliable than the others and, being Genoese, he was not affected by local nepotism. The two men understood each other immediately. Profumo was clear and sober, but had to tell Dohrn that he understood nothing about the technique of an aquarium. Did Dohrn understand enough of it? Who could have replaced Kelterborn? There was no one. But Lloyd was still in England after building a new aquarium in the Crystal Palace in Sydenham. Lloyd back in England! Dohrn turned to him, asking for help and advice; and Lloyd agreed with friendly enthusiasm although he was told that no fee could be given to him. A complicated situation arose – he became so to speak »corresponding building director« for the technical installation. Dohrn had, of course, acquired knowledge in these matters, but he was after all not an engineer. A vast correspondence, also about individual questions, went back and forth. It was written in English and Dohrn himself had to be the interpreter to Profumo and his coworkers.

There was trouble in the Italy of that day with the continuity of local administrations. Naples was considered the most »turbulent city«, with its changes of majorities and therefore of mayors. It was more or less up in the air which Principe or Duca, Marchese or Conte* was to be elected by the preference of the people, as well as whether he understood anything about finances and administration. If things came to a crisis, the Roman government would dismiss the administration and, for the time before a new election, send a Royal Commissioner. Count Capitelli, with whom Dohrn had drafted the contract in the spring of 1870, had made way, a few weeks later, to a Royal Commissioner. Since the spring of 1871 there had been a so-called orderly Municipio, but the questions of site and building permit stopped completely. There was still the project of the Belgian to carry out, the large new road along the beach – to whom did the beach belong? To the State or the City? The legal suit went on for months, until they dug up an old royal privilege: the beach belonged to the City. As long as this went on, the City Hall did not really want to speak with Dohrn. The enormous street

* Principe = Prince, Duca = Duke, Marquese = Marquis, Conte = Count.

project was much more important. Now that matters were clarified, the short-lived Municipio was again removed. Would the Government Commissioner make permanent decisions? This time the new confusion brought luck. Rome had placed a professor of mathematics, Trudi, at the head of the city, and he wanted to render a service of science. Although he did not consider himself justified to write the final form of the contract, he took care of the matter. He gave them a new site which he considered to be more favorable – it is the one where the Station now stands – and above all he gave permission to start building. For Dohrn this was the most important document.

Dohrn wanted the builder, Profumo, to start building, and was not worried about the fact that this beginning without a contract might become dangerous. He himself wanted to go North. He thought he was not needed and was anxious to finish his work on pike and perch, which he had left behind in Stettin. Thus, he was not present when the workers started in March to dig up the ground. But a few weeks later an old traveler in Italy walked around the wooden fence, somewhat curious and somewhat proud, with divided feelings because, after all, his own money was being put into the ground. Carl August Dohrn wrote casually to Anton that matters had obviously become serious.

Five weeks later father and son met in Florence. The father was sick, but, as was his way, he did not want to admit it and refused to see a doctor. The matter did not seem harmless, so Anton asked to accompany him to the Spa of Lucca, which he did not know. There, the 66-year-old man became sicker, but he fought stubbornly against having a doctor or medical care despite his high fever. This caused an odd situation: Anton asked the host to give him, secretly, a room next to his father's in order to be present in case of an emergency – his quarters had been in a different place. Now it happened that one morning the old man, confused by fever, entered his son's room, perplexed, amazed, and finally moved, when he heard that Anton had slept in that room for days. »This I will not forget, son!« This anecdote belongs to the history of the

Station. Carl August, in his own naive egocentricity and in the awareness that he may have judged his son wrongly, doubled Anton's credit after this event. This was very important at that moment, since more expenses were constantly accruing.

In Naples, where he returned in May 1872 after two month's absence, new trouble awaited him. The walls were going up fast, but on the wrong level. Before leaving he had talked over the ground plans and transverse sections and had expressed his worry over whether the plan for the basement vaults was in correct relation to sea-level. The architect referred to the mesurements of the Municipio. Now it turned out that he had indeed proceeded in error and had figured the foundation too high by one meter. That was precarious because the permitted measurements of the entire building were displaced. The architect calmed himself and Dohrn, saying that nobody would notice it. Profumo, too, thought that nobody would worry about it later. Moreover, the relationship between the Neapolitan and the Genoese had become unpleasant. The former was constantly late in delivering the plans. Profumo tried to hurry him; there was no end of friction and unpleasantness. An open conflict broke out between the two. Dohrn had to make a choice, and he opted for Profumo. There was a somewhat frightening theatrical conclusion. The architect, who took the plans with him, sued Dohrn because he had hurt him professionally, he had lost wages, and his health was affected. He asked for 40 000 francs – a lawyer had written all of this out quite efficiently. Fortunately, at this moment at Profumo's house there was a well-versed Neapolitan lawyer, to whom the worried Dohrn went immediately. He simply laughed and dictated a counter-suit to Dohrn (who quickly calmed down), detailing stolen plans, lost time, missed terms – the total of these sums amounted to exactly 40 000 francs. There was of course no court action. The architect withdrew. Dohrn forfeited only the contractual sum of 700 francs.

The consequences of this bad relationship were to become apparent later. For the building itself, the solution was positive. By accident, Adolf Hildebrand happened to be in Naples during that period, and he was delighted to take over the planning for the façade, he cut out the decorative playfulness of his predecessor, and gave it the neat, harmonious measurements which Dohrn had indicated in his earlier sketches. The rhythm of the loggias and the kind of pillars and capitals which became the models for all later enlargements were decided on by the young architect, who stood at the beginning of a great career.

6 THE PROGRAMMATIC ESSAY

Dohrn's planned activities as a scientist could, in these weeks and months of running around and negotiating, not amount to much. But Dohrn had the fortunate idea of writing a condensation of the things which had moved him for years and to which he was determined to devote his life. He hoped in this way to reach the interest of the educated classes. It was not a question of publicizing his financial worries, which gradually became very troubling – that might have been fatal. He had to find a form which would make the general meaning of the enterprise clear to the layman. During the summer of 1871, while preaching to his wealthy friends, he had acquired sufficient understanding of the difficulty in demonstrating the elementary quality of his plan. It was not only the question of a zoological museum or »something like it«, nor a display for curious or bored people, or even for those eager to learn something; rather, it was something completely new, a work place for systematic, continuous studies. The concept and nature of a »Research Institute« did not exist at that time. Upon the initiative of a private person, an example was to be created, which later should have a fruitful following. If one could show this as a new vision of zoology in general, and also of its self-awareness, at the same time influence could be gained in the official or corporate offices responsible for the cultivation of the sciences. The personal side of the attempt would almost entirely disappear behind the proof of objective necessities, Dohrn himself would appear only as the person who fulfilled a task that was set at this period by the state of science itself. Looking back, the entire enterprise would appear as a bold and grand precursor of the numerous independent institutes which arose a few decades later, dedicated to the individual sciences. From the point of view of the history of the mind, it introduced to the intellectual consciousness the inclusion of natural sciences as equal components in education.

In June, 1872 Dohrn wrote his extensive essay: *The Present State of Zoology and the Foundation of Zoological Stations.*[21] It has the character of a memorial. The national-liberal journalist and representative Wehrenpfennig took care that it was published quickly in Treitschke's *Preussische Jahrbücher,* the most important German journal of that time. This change of place of publication of the former admirer of Johann Jacoby seems characteristic. From this place his word reached the educated German bourgeoisie and the open-minded officials. If this goal could be reached, it might also be useful to give to the Italians an idea of what was going to happen on their own coast. Luckily, a translation of the essay could be published almost simultaneously in the excellent Italian Review, *Nuova Antologia.*

This great essay is considered by some zoologists as among the »classical writings« of their science. It is indeed a masterpiece of writing, a keen and clear presentation. According to it, taxonomy alone, i.e., the collection, description, and naming of faunistic phenomena, does not advance the knowledge of the laws and richness of life. Dohrn connects this with the theses about a new type of problem for future scientific research. Others had already seen this. Dohrn does not appear with the gestures of a revolutionary reformer. He mentions the work of Karl Ernst von Baer, the fortunate individual investigations of Siebold, Leuckart, and Weismann, who illuminated parts of animal development. But, following Darwin's appearance, these important beginnings needed a methodical enlargement and a new safeguard in the entire activity of teaching and research. In Dohrn's words: »The program of zoological research for the next century seems to be expressed in two thoughts:

(1) the fight for existence and the consequent natural survival of the fittest; (2) the recapitulation of the development of the species through the history of the development of the individual.«

Dohrn is clever enough not to make his own plans central. In a light, not solemn tone, which yet does not avoid self-irony, and courageously offers the comparison with Don Quixote, he adds autobiographical details, talks about his attempts in Scotland and Messina, his hopes and failures. A larger space is devoted to suggestions regarding the state of science at German universities: the teaching appointments should be separated and doubled; next to the comparative anatomists there should be an observing biologist trained in physiology; with limited means laboratories should be created everywhere, with small aquariums and artificial pools, in order to relieve researchers of the efforts and accident of finding material. How much knowledge is lost in the traditional way of handling zoological gardens. The English, surpassed by the Germans in the breadth of research energy, are more practical and systematic, whereas the Germans incur the danger of overestimating museum collections, not understanding how sterile this has become (here we might hear a slight polemic against Dohrn's Berlin teacher, Peters, who collected and mummified specimens).

For the creation of a thorough taxonomy it had been indispensable to use nonprofessional but enthusiastic men such as country teachers and vicars. Their participation, which had diminished for a number of reasons, had to be regained by showing them the urgency of the new goal. It was no longer important to determine the species or group to which an animal belongs, although of course taxonomy remained the presupposition of each learned conversation or work to protect it from misunderstanding. Now it was a question of observing ways of life and stages of development, and of observing the pairing, food, climatic reactions, etc. Science could not do without the willing cooperation of the educated nonprofessionals in this new method. Zoology now had before it the possibility of stepping from the periphery of interest to the center. This was the result of the feuds around Darwin. It was, of course, the task of the zoologists to realize this new chance by convincing work, to ease the way of young, talented, waiting scientists financially, so that the choice of this profession would no longer be reserved for the rich – the public consciousness or at least the good will of those responsible for the intellectual situation of the nation must cooperate.

Dohrn's own scientific starting point had been comparative anatomy, but this no longer satisfied his morphological interest. Fertile morphological work seemed to him possible only on an embryological basis, and this insight determined his own research. This programmatical work aimed further: »Nevertheless, there is something missing, which is at present clearly felt, namely that we have very little to report on the way of life of most animals whose external and internal anatomy we know by heart.« In order to understand »the natural selection« of animals one must learn the factors of the fight for existence, the habits of life. »Therefore, exact statements on the place of abode, on food, length of life, pairing and length of pregnancy, on relationship to the eggs or the young – in other words, everything that concerns the isolated existence of an animal, have the greatest scientific value. Even more important will be the knowledge of the living together of animals . . .«

These sentences indicate clearly enough the central concern, namely, to create a place for the real and far-reaching science of life. In this sense, he took the basically important step of demanding the renovation and broadening of physiology, »the creation of comparative physiology or rather of a physiology which observes and explains not only the existence of bodily functions, but also their coming into existence«. The physiology taught at the universities, which was mainly interested in nerves and digestion, has become more and more a branch of medicine. It leaves to morphology, which was already overburdened, »the fields of the physiology of procreation and growth, furthermore the problems of motion and the organization dependent on motion, actually the entire field of correlations within the animal organism . . .« Physiology had not yet understood with what amount of new questions it was being burdened by the theory of evolution: »It will be the foremost task of comparative physiology to use all its methods, all its subordinate sciences – in this case chemistry and physics – to look for the functions of all animals, even the lowest, and to state their origin with the help of embryology and comparative anatomy.«

Whether or not we call the program of this great treatise prophetic is unimportant – it must be seen as a judgment on the scientific disciplines from the point of view of the year 1872. In an outstanding man like Johannes Müller the elements of comparative anatomy and physiology were combined. In scientific research, however, they were separated (not only in teaching, where this was unavoidable). The biographic importance of the great treatise is the fact that he not only visualized, theoretically, the various groups of sciences as one, by pointing out to them neighboring or common tasks, but also that he gave reality to what he had earlier simply described. He did this with unerring consistency by forcing the situation, or by patiently letting things mature, or as an organizer who made the possibilities real, even if only after decades.

7 THE BUILDING CRISIS

In the summer of 1872 the Station grew under the direction of Profumo and according to Hildebrand's design. Dohrn could go to England and Germany to try to find new possibilities for financial support. He also had new ideas to report. But a wire from Naples tore him away from promising negotiations; the city administration had stopped the continuation of the building and made him responsible, because the height of the building, as indicated in the contract, was being surpassed. The bold opinion that the mistake would not be noticed and Profumo's belief that nobody would care had been wrong. Dohrn found himself, because of the change of personnel, facing a new Royal Commissioner, Marvasi, who was a highly esteemed jurist. He said that a contract was a contract, he could not change it; in three months a new Municipio was to be elected – perhaps Dohrn could get permission from it. Should he lose those three months? Supported by Profumo, he decided to let the building be continued and to ask the new Municipio for indemnity.

That was a dangerous game. As the walls grew above the fence and the large arches of the loggia at the top floor became visible, the imagination of the Neapolitans took over. What was this to be? A café, a theatre? Or the private palace of a foreigner who would be allowed to settle in the most beautiful park of Naples? Dohrn could sense that the curiosity had distrustful and negative components. There was a friendly comment from a paper vendor which Dohrn heard in passing: »At the bottom there will be a fish market, upstairs a small university.«

The next election had an unexpected result: neither the Moderates, with whom Dohrn had dealt up to now, nor the Radicals, who had followed the recommendations given by Garibaldi's friends, gained control, but the clerics, or rather the old bourbonic aristocracy. In one of the first sessions of the city government the question was settled – building was prohibited. The only thing Dohrn achieved was permission to continue building the flat roof (required by Hildebrand's changes) in order to prevent the harm that would be done to the inside. That was all. The report of that public session was condensed to the rumor that the new administration intended to tear down the building that had been accepted by their precursors. This was partly meant as a gesture toward arrogant foreigners who believed they could occupy the most beautiful site of the city. Dohrn ran from Pontius to Pilate and from Pilate to Pontius. He could no longer become furious, because he was faced with a hostile administration. Formally, he was of course wrong, but Naples would one day be grateful for the entire enterprise. On some of the people he believed he had made an impression, but that remained vague. Fate seemed to rise against him. A big flood, which also destroyed the Molo, broke into the building on December 5, tore away the building shed, washed out the lime pits. The few violent hours did 3000 to 4000 francs' worth of damage. And before they had repaired the trouble, the city hall sent its decision retracting the provisional permission to finish the roof. Dohrn was told that the mayor himself, Count Spinelli, was his real opponent; the Station spoiled the view over Capri from his own place. The matter was entirely hopeless. The nervous tension of Dohrn became worse due to a letter by du Bois-Reymond, which arrived on December 24. The helpful man had presented the matter in Berlin to the Academy of Science, optimistically and eagerly. Now the letter indicated that »dangerous weapons« were being prepared against Dohrn; and he should please look somewhere else for help. On the same day, Dohrn went to Berlin to see what these weapons were. The improvised trip allowed him to put things into some degree of order. From the alarming uncertainty concerning the support from Germany, a better sense of security developed in the few weeks to which the trip was extended. But the impression which the German Consul General Stolte gave him upon his return to Naples at the end of January was all the worse. The neighbors of the site had formed an alliance to force the destruction of the building. All of this because of the mistaken calculation of the level by Signor Capocci – at least that was their pretense.

Did these people know what risk this was? That it was more than the whim of a foreigner? It was a blessing that he had succeeded in having the *Programmatic Essay* printed in the preceding summer in the *Nuova Antologia*. Of course, who could know whether it was in the hands of the decisive people, whether they had read it, whether they would read it if he sent them the issue now? With little hope, but in the interests of not missing a single opportunity, Dohrn sent the issue to some of the important men among the administrative and political leaders. The echo was not encouraging. The Prefect Mondini told him that, although Rome wanted him to support Dohrn, he could not do so. It was entirely a city matter. As a private person, he shared the judgment of the city administration and public opinion.

It was one of the first days in February. In the morning Dohrn visited the German Consul General Stolte and received from him a note which had just arrived:

Yesterday I received an essay from Mr. Dohrn, which I read with great interest. Please tell me where this young man lives who, in our time, has the courage to leave his homeland and to risk his time and money in order to create a scientific enterprise. I wish to look him up and help him.

Yours, Giacomo Savarese

The author of the letter was the majority leader of the Municipal Council. He had played an influential role in the public life of Naples. Once, under the second to last Bourbon, Ferdinando, he had been a member of the liberal ministry which the latter had to call together. As Dohrn characterized him: »Savarese impressed me as a noble man who, by conviction and desire, was more liberal than most of the party members who called themselves liberal. Simultaneously, he was convinced that the Italian people needed a tangible cult such as only the Catholic Church could offer.«

Of course, Dohrn did not wait on this promising morning for the old man to visit him on the fifth floor of the Palazzo Torlonia (with its 185 steps), but went to see him and was received immediately. »Yesterday I read your treatise and I know you through it. I also know that you have to fight with difficulties. But do not worry. I will know how to defend you.« Dohrn suddenly felt »transported into a different atmosphere«. Here was a man who did not take him for a speculator but rather understood what his purpose was, and who did not withdraw into friendly declarations of sympathy but who was willing to fight for him. He stayed with the old man, who had received him in a dark silk housecoat, for almost three hours. They talked about Naples, Goethe, and Kant; and the baron was amazed that one could love and honor these authors while still adhering to Darwin. This gave Dohrn the opportunity to clear up partial misunderstandings. They talked about philosophic differences, not growing apart, but rather meeting on a human level. Savarese became very friendly; he would take care »that no stone would be removed from the building.« And when, two weeks later, there was a violent fight about the Zoological Station and its height, which violated the contract, Savarese and his friend Prince Torella appeared as defenders. Behind them a considerable majority came together, against all prediction. That meant more than a formal victory. Savarese's appearance gave to Dohrn the much-needed moral support in the public consciousness of the city. In an almost desperate situation, a deus ex machina had descended, called up by the words of deep conviction. The campaign for science, which was only in its early development, would still live through many battles, hostile encounters (either evaded or carried through), defeats, fights for positions and victories, surprise attacks and fatigue. The three years full of enormous psychological and physical strain had been a tactical lesson for Dohrn. They had partly wiped out his reckless naivity, but had also strengthened his power of achievement and a certain stubbornness to force the ultimate success, despite depressions that had to be suppressed or to be allowed to take their course.

8 THE ROAD TO THE REICH AND THE CONFEDERATE STATES

The history of the building had to be told in its essential features. It is not only an external event. Rather, it gives a picture of the adventures which the young man encountered; it also gives a picture of his nature, of his ability to make decisions or to evade them, and of his imagination and his patience in difficult times.

The steps for the inner plans were more decisive – Dohrn had, in his plans, taken on too much financially. The increased credit from his father was quite ample, but it had to pay not only for his own living expenses, but for Grant's and the first assistants' as well, whom Dohrn had already engaged – he was not made for loneliness. The various incidents and interruptions, indemnifications, and destructions spoiled his tight calculations. The popular opinion which surrounded him in Naples endangered his optimistic calculation that the entrance fees to the Aquarium could carry the current scientific work. The planned extent of the inner organization and preparation of work spaces had grown – his imagination too easily anticipated the desired development as having already been achieved. Who was to populate the rooms, even if it were possible to find the capital for the finishing touches? That, too, was doubtful.

Should he simply go to the State and ask for money? And to which State? The decision was difficult. If he did not succeed, it would also be a moral defeat. If he was successful, he might endanger the valuable freedom needed as a »motor of life« by such an enterprise without comparison or predecessor. But he did not come with empty hands: he was offering to science an entirely novel chance for research. Who were the advisers of this science? A colorful picture showed itself: states with ministries and universities, learned societies with or without money, foundations, perhaps, mainly in the Anglo-Saxon region, rich private people.

In the spring of 1872, in the middle of troubles with the excitable and unreliable architect, Dohrn had an idea which became most useful for the entire future of the Station and the development of science toward an international symbiosis: the system of the so-called work tables. Dohrn described the road to this idea: »I had heard that corporations or rich private people donated ›beds‹ in order to procure money for the creation and maintenance of hospitals. They gave either a yearly definite sum or a larger sum à fonds perdu to the hospital and preserved by this the right to dispose of a ›bed‹, i.e., a place in this hospital for the care of a sick person: »This served me as a model for renting various work places (which were to be established in the laboratories from the beginning) to various governments and scientific corporations. This would guarantee a certain yearly income which would allow me to borrow more capital.« That was the answer. In the following years it became Dohrn's main task to manipulate this method with great virtuosity.

The new plan was to be communicated first to his English friends. In a quick trip – that meant at that time 81 hours – Dohrn traveled in the middle of August from Naples directly to Brighton, where the British Association was holding its annual meeting. He did not succeed in his attempt to win this society to rent a table immediately, but a committee was nominated to discuss the case. The physiologist M. Foster, who was on his side, thought he could pledge his word firmly for the University of Cambridge. Of course, many technical things were discussed in London with Lloyd. Above all, he brought home the promise of scientists and scientific societies to send the books they edited to the scientific library to be created. Dohrn's own rather extensive scientific library had already found a provisional space in Naples since December, 1871. He had donated it as the basis of a collection which, in the course of decades, would be made complete and unique throughout the world.

In early September 1872, Dohrn went home. The programmatic essay from the *Preussische Jahrbücher* (it had been published in June) he had sent to the President of the Office of the Prime Minister, Rudolf von Delbrück, and to the recently instituted Prussian Secretary of Education, Adalbert Falk; these two men seemed to him the most important with regard to his field. After a written question, Delbrück was willing to see the young man on September 10. Twenty years later, Dohrn related this conversation in the *Deutsche Rundschau.* He reported his plans and needs to the immovable face of the Prime Minister, and closed with the question whether the government of the Reich* could support him with 10 000 Thaler. »A dry ›No‹ was all he answered, whereupon I kept silent. However, I remained seated since the Minister did not get up. But my thoughts were very active. For the knowledge that this moment was of decisive significance was lively in my thoughts. Finally, I started to speak again: ›Your answer, your Excellency, destroys my hopes in a single stroke. I may honestly claim that no selfish mo-

 * Germany's individual states were united into one German Reich in 1871.

tives guided me, but that I am trying, with a clear conscience, to work for the interests of science, particularly German science. I have risked my money, my career, my freedom, and have shown that I am not afraid of difficulties. And now, when I hope to be supported by my own country in this endeavor, Your Excellency turns me away.‹ The minister answered: ›It is a doubtful matter to which the Reich cannot contribute.‹ Then I answered: ›Allow me, your Excellency, to tell you of a little event that happened in September, 1870. During the war I was in a garrison in Kassel. After the battle of Sedan* I had furlough for a short trip to England where I visited the meeting of scientists in order to promote my enterprise. In London a literary friend invited me to dinner; there were other scientists present. There was of course talk about the amazing German triumphs. Suddenly one gentleman said: ›One consequence of your German victories will be the fact that your scientific superiority will be broken. Now you will submit to militarism, and one consequence will be a slowing down of science. Up to now we have translated your textbooks, now you will have to follow our leadership.‹ I kept silent at this odd assumption. But I promised myself that I would do what I as a single person could do to annihilate this prophecy. But if Your Excellency now speaks of a sloping ramp, I am particularly sorry that my efforts were in vain.« During my account there was a movement in the features of the Minister, and when I had finished, he said: ›While I do not say, 'absolutely not', I can equally not say 'absolutely yes'. With a sigh of relief I said immediately: ›On what does the decision rest, Excellency‹? The Minister answered: ›Above all I would have to have an expert opinion in my hands, for example from the Academy of Science, saying that your enterprise is worth being supported by the government of the Reich. If you give me such an expert opinion, I could consider the matter more closely‹. I felt sure that the Academy would give such an opinion under all circumstances, and

* This was the decisive battle in the Franco-Prussian war in which Prussia was victorious.

I took my leave.« Dohrn's confidant in the circles of the Academy was the physiologist Emil du Bois-Reymond, a naturally pleasant character, always willing to help, with a slight tendency toward vanity, but also always ready to recognize good will and achievement. He readily undertook to bring to a quick and correct execution a request addressed to the Academy. Dohrn could be sure about the eagerness of the busy scientist who was also professionally interested in the Naples enterprise. Also a visit to the new Secretary of Education, Falk, went well. Falk had been in that office for only a short time, and his task was to lead the fight against political Catholicism. In academic circles his name was still unknown. He welcomed the opportunity to demonstrate his independent desire to be supportive. He talked about these tendencies in a lively way. Dohrn suggested that Prussia should »rent two work tables«. The Secretary promised his support. Everything went well and the two discussions with the Ministers offered a good foundation and the idea of investigating private possibilities – this was the moment when the unpleasant wire from Naples brought news of the interruption of construction and called Dohrn back in a hurry to the chaos in Naples.

Du Bois-Reymond felt fairly secure. The reports he gave demanded a little patience, because first of all a commission had to be formed, but they sounded optimistic. But then came the alarming letter to Dohrn which said that there existed a »less favorable mood« in the mind of some academicians; in the best case they could count on only a lukewarm recommendation. When he had recently written about the »good prospects« he had had no idea of »the dangerous weapons which, if some one wanted to use them, existed for everyone to use against you«. They were of a kind which might be used perhaps even outside the academic world. Dohrn »should please not become depressed, but should strain every nerve to get the necessary help from other places«. That meant that Dohrn's Berlin supporter considered the battle lost. What had happened? He had to clarify the situation. The day he received the letter, December 24, he immediately started his trip. It was the most unsuccessful Christmas of his life. He used his stay in Munich to visit C. Th. E. von Siebold. The old gentleman was one of the friends of Carl August Dohrn and had for years sympathized lovingly with the son. Now, lying sick in bed, he grumbled so refreshingly against the Berliners, particularly against his close zoological colleague, Peters, that Dohrn felt much strenghtened. He also gave him a few recommendations for Count Usedom, the director of the Berlin museums, who was an intimate friend of the Prussian Crown Prince and his wife. Dohrn had to promise him that upon his return he would stay in Munich somewhat longer – he would at that time introduce him to the great men of the University, namely Liebig and the botanist Naegeli. This interruption of his trip later proved well worthwhile.

Du Bois-Reymond was, of course, somewhat astonished when, the next day, expecting at best a complaining letter, he saw in front of his door the receiver of the letter in an impatient, fighting mood. »What is the matter?« The academicians, at least in part, had their worries: in the last few years Dohrn had not produced any scientific publications which would testify to the state of his knowledge. Dohrn had also mentioned the expected income from the visitors of the aquarium – obviously this was a business deal, and even giving it moral support was beneath the dignity of the Prussian Academy (the protocol of the Academy of December 16 speaks of »the intended plan as being mainly an industrial enterprise«). Peters had told everyone that the young man had, almost seven years ago, not passed the doctoral examination with him. In their conversation some renewed explaining and arguing occurred, which examined and clarified the individual objections. Du Bois-Reymond was himself infected by Dohrn's dauntless will to fight. Dohrn decided to visit in person his enemies Beyrich and Braun, Pringsheim and Peters, as a man in uniform who wanted to spend Christmas with his family and, on that occasion, inform himself of the state of affairs. Peters was somewhat embarrassed by the renewed acquaintance. He did not mention the exam; he believed that 10 000 Thaler per year was a little too much. Dohrn did not quite know whether

»per year« was an excuse or an honest misunderstanding. Of course, Peters said, after being informed about the one-time donation, one could talk about that. Similar positions were taken by the others. Secretary Falk, whom Dohrn informed about the difficulties, said that the Department was not interested in details, it only wanted a word on the scientific usefulness of the enterprise. Now they talked about three Prussian work tables – but for that, one would have to win over the Secretary of Finance.

Altogether the forced trip to Berlin proved fruitful. Count Usedom, who had been Ambassador to Italy, could evaluate better than the others what Dohrn had already achieved; he was impressed by it. He offered to see to it that a copy of the *Preussische Jahrbücher* reached Crown Prince Friedrich Wilhelm. If Dohrn were under the »protection« of the Crown Prince, that would be an advantage for the position of the Secretary of Finance. And perhaps one could appeal to the friendship of the Prussian and the Italian Crown Prince (Umberto). The perspective opened up by this conversation with the educated and experienced man was promising. Friedrich Wilhelm was quickly ready to express to Falk his personal interest in Dohrn's case. At the same time, he had asked the German representative in Rome to work »in a decent way« against »the difficulties already caused or still planned«. Even Delbrück was no longer the cool official, the atmosphere seemed changed, the Minister listened to the Naples stories, laughed about the demands of the first contract and was glad that Dohrn had active helpers even in Munich. He only shrugged his shoulders about the worries of the academicians. The Station should not be painted »black and white«, but »black-white-red«. * Dohrn felt that the relationship had left behind the merely official character and had become personal.

Checking up in Leipzig, on his way home, he found that the efforts of the zoologist Leuckart had met with the distrust of the

---

* These are the colors of the newly founded German Reich. The Prussian colors are black and white.

physiologist Ludwig. The latter declared that he was not competent and would therefore give a negative vote. There would be trouble to win Saxony. The situation in Munich proved more positive: Above all, Dohrn received a personal letter from Liebig to the Italian Minister of Finance, Sella, who, being himself a natural scientist and an important man, helped quickly. Sella could not allow him exemption from customs and taxes for the imported instruments, but he supported the readiness of the Minister of Education, Scialoja, in taking two work tables for Italy.

Those four weeks had been hard, and Dohrn returned to Naples at the end of January in a state of exhaustion. This was the moment when the propaganda for the destruction of the entire building had acquired threatening proportions. Fortunately, a new letter arrived from Berlin. The kind du Bois-Reymond, who remembered what he had sent to Dohrn as a Christmas present, noted on the envelope »good news« so that his handwriting would not appear as a harbinger of disaster: the battle had been won due to Dohrn's personal appearance, and the professional recommendation demanded by the Ministry had been accepted by the entire Academy. A few weeks later news came from Delbrück that now 4000 Thaler would be sent and 3000 each for the following two years. His home land had not forsaken him. In Naples, Baron Savarese had just saved the situation; but it was by no means secure.

9 FRIENDS AND COLLABORATORS IN THE EARLY NAPLES PERIOD

The boat of life with which Anton Dohrn meant to reach new coasts soon acquired strange passengers of various origins. Some left the boat after some time, others found a life together with the Station. And one cannot think of them apart from the Station's strenuous growth and proud fulfillment.

The first person Dohrn called to his side was his friend of Jena days, the English writer Charles Grant.[22] Considered rationally, that was a surprising decision. Grant understood nothing of the scientific or business problems and the country was foreign to him. Since Dohrn knew about his complete lack of money, he was voluntarily taking on a new responsibility even though he was acquainted with his friend's frugality. This caused him some worries. But his lively nature needed a good listener and a man who, in times of depression, remained indulgent, patient, and helpful. Grant became a kind of side figure in the early history of the Zoological Station. Without ever taking over a definite function or even a real job, he was available, helped out, reported events to the traveling Dohrn. He had, with a certain passion, taken part in Naples' daily life, as an attentive, quiet observer, full of tact and love. He became completely indispensable as a mediator in innumerable everyday occurrences.

Grant was also a person not easily understood in the life history of Hans von Marées and Adolf Hildebrand. He had been Hildebrand's teacher in Jena at the Stoy Private School. As a young English student, son of a colonial merchant who died young, Grant had gone to Jena to improve his German. He wanted to become a teacher of German later in his homeland. But his heart was captivated by Germany. He found friends there, and became very familiar with German national customs by taking long hikes through Germany. In the same way in which he tried to interpret to Germans the development of recent English poetry, he also wanted later to explain the German mind to his English compatriots. Huxley, as we saw earlier, was supposed to see whether a British publishing house would publish Grant's translation of Kleist. Marées painted Grant several times. Dohrn, in an obituary, described his friend after his death, remembering the Jena days: »A large, broad head full of black hair which was not well-combed, very expressive, sparkling eyes, a reddish face, a full, not well-proportioned upper body, on legs that were too thin, in extremely neglected clothing – thus Grant moved on walks among the boys.« The friendly relationship had started in 1866. Despite political disagreements (Grant supported Austria, for which Dohrn wished defeat) they sympathized in literary and intellectual matters. At this point Dohrn needed just such conversations, stimulations, understanding, since he was in danger of being forcibly swallowed up by business and financial worries. Later he expressed it in this way: »The new profession would have made the usual fields of my intellectual activities so desolate, that I feared becoming bankrupt unless I always remained close to the natural sources of my intellectual and emotional life through daily discourse with Grant.« Grant, a few months younger than Dohrn, had lived in Berlin since 1869. There he tried, without much success, to write for English and German journals. Now he followed the call of his worried friend and stayed in Naples for almost ten years, interrupted only when he visited Adolf Hildebrand in Florence. The way to Naples was an odd detour for him, to rediscover the relationship with his homeland. His friendship with the young English scientists in Naples opened up some doors to important British journals. He hated Gladstone and believed in Lord Salisbury.* His poetic productions were few – a little volume of lyrics and an attempt to dramatize the *Edda,* which shows that he lacked the inner tension needed for a dramatic writer. But after his death, a collection of *Stories of Naples and the Camorra*** was published. Its German translation won the admiration of Hermann Grimm. More clearly than his essays, this book shows a great facility in the art of representation. As a manner of story-telling, it is a classical description of Neapolitan folk life. If we did not know from his correspondence how deeply Grant was involved in Dohrn's local negotiations, this

* William Ewart Gladstone (1809–1898). Liberal prime minister of Britain 1868–1874, and again 1880–1885. Created the elementary school system. Being Scottish, he tried to help Ireland, but overlooked the Protestant part around Ulster. His politicial opponent was Cecil, Third Marquess of Salisbury (1830–1903). British statesman. Became Foreign Secretary in 1874 under Disraeli and Prime Minister 1885, following Gladstone. Defeated Gladstone concerning the establishment of Home Rule for Ireland. Remained Prime Minister until 1902.

** Camorra was a secret society, similar to the Mafia, based in Naples.

book would show him to be an authority on problems of counseling, as far as the psychology of his environment was concerned.

In his last years, Grant lived and traveled in Carinthia and Styria.* But he had become an alcoholic, and many of Dohrn's moving letters told him in which institution and with which doctor a cure might be found. He shared his worries with Adolf Hildebrand. Grant died in Graz in 1889. None of Dohrn's contemporaries was as close a friend to him as Grant had been.

To the core of the colony which had been transplanted from the idyllic Thuringian town of Jena to restless Naples belonged the Baltic zoologist, Nikolaus Kleinenberg. Grant was kind, softhearted, leaning toward Catholicism. Kleinenberg, by contrast, was a cool skeptic, not without arrogance, and often painfully harsh. Dohrn had met him in Jena. The word »intimate« must be avoided in describing relations between the two men, since Kleinenberg was too conscious of the difference between them and remained aloof from his colleague, in a kind of defensively expectant position. Basically, Dohrn was the one who wooed him. He greatly respected Kleinenberg's efficacy in his field and his original approach to attacking problems. His fascinating irony, his love of paradox, had for Dohrn the charm of intellectual training. They shared a common opposition to the dogmatism of Gegenbaur and to the blossoming self-assertion with which Haeckel interpreted Darwinism as a general philosophy. Such amiable popularization with its wide success was an easy victim for Kleinenberg's wit and dialectical way of thinking. The portrait Marées painted of him shows a narrow skull with a cold, somewhat bitter intelligence, and a sharp sense of observation. Financially he was in bad shape, with an eternal lack of money which he had no expectation of changing, but he was too proud to look for support. Dohrn wanted to help him and to help himself at the same time: if Kleinenberg would take over the directorship of the Laboratory, Dohrn could be certain of the scientific level and demand. He saw administrative obligations

* Kärnten and Steiermark, two Austrian provinces.

approaching him swiftly, and had already, during his first stay in Naples in 1870, made firm suggestions to Kleinenberg: »I would like to win your strength now for the new turn our science will take.« He courted him the following winter but received no answer; he kept him informed on worries and successes; he even missed him on a human level, despite the enjoyable presence of the young Ray Lankester: »As you know, I count less on your cavalier reputation, but I feel in you the mature man, who lives maturely and energetically whatever his goal . . .«

The experiment was unsuccessful. As long as the development was still partly chaotic, Kleinenberg was an enlivening element, but he failed when matters became orderly and a realm of responsibilities was created. As a trivial example, he was to have the responsibility for the »tables«, the caring and counseling; that started around 8 a.m., but Kleinenberg was indifferent and lazy. He appeared when it suited him, and there were troubles and complaints which endangered the new and not always positive reputation of the Institute. The break came in 1875 due to the harshness of the man who could neither accept discipline nor bear dependency. »Your justice is generosity, my generosity is justice,« Kleinenberg had once written. But this paradox does not completely portray the inner situation, although it indicated the stronger rigorousness of Kleinenberg.

Kleinenberg and Marées had become friends, and Dohrn later asked the painter to mediate, but he rejected the plea. Asked for advice, Dohrn several times recommended his old friend for appointments. He did not hide the difficulties of his nature, but strongly emphasized his scientific qualities. This was successful in Messina in 1879. Then Dohrn received a characteristic letter: »From the report of several friends, I understand that in the appointment for the open position in Messina, your participation in my favor far surpassed the limits I could expect and which I could not accept without writing you. It is the purpose of these few words to be my thanks, and you will not miss the intention if you take my thanks as I give them: quite free and sincerely.« This note broke the silence, »since implacability is not in my nature«, Dohrn answered. But due to his past experience, he left the initiative of writing to Kleinenberg. The correspondence continued to a certain degree, but remained at a diplomatic distance. Kleinenberg died in 1897; in the history of zoology he is known for his writings about the Hydra.

The third man who appeared in Naples was very different. Dohrn was somewhat disturbed when, in November of 1871, he received a letter from Dr. Hugo Eisig. Eisig had read a newspaper article in which Carl Vogt announced to his readers the intended foundation of his young friend. Eisig offered his services. He had no academic expectations and no money, but he did not need much; at that moment he was surviving as a writer. Wherever Dohrn's steamer went, his own little lifeboat could be tied to it. From the Jena period they had known each other personally. Dohrn was uncertain. He would not have picked him as a life companion. But who could know whether the man would perhaps be useful as a second assistant? Their common friend, the geographer Fritz Ratzel, had written a letter in which he praised Eisig's character in strong words. Dohrn wrote to Kleinenberg about Eisig's application and his answer: »I scared him by saying that I had become unbearable, despotic, full of whims, lazy and selfish, in order to scare him off. But he will probably say: that's what I want – it leaves me without responsibility.« And Eisig wanted it. He came, while Kleinenberg still avoided him. Dohrn wrote him in June, 1872: »Eisig is here, a rather boring man but so intelligent and decent. But he begins to understand that he may learn a lot from us, although in the beginning he covered me, like a minister, with his empty wisdom. The main thing is that he is diligent, of good will, and he understands the heaviness and difficulty of my task.«

It was not easy for Eisig to create for himself a human position, when the Naples circle, so rich in intimacy and tension, grew wider, particularly through the addition of the artists. Kleinenberg was hard on him and offended his sensibility. But once the machinery had started to run, Dohrn found in him an element of reliance which he needed badly. As a representative in negotiations he was not quite sufficient; he lacked imagination and security, mainly in the beginning. But he did not neglect anything. When Dohrn returned from a longish trip in 1874, he wrote to Kleinenberg: »The Station is in very good condition. Like a good housewife, Eisig has visited untiringly every corner in which disorder and lack of cleanliness might settle and has driven them out thoroughly. I solemnly gave him the testimony that he had served the Institute and all of us very well.«

Shortly before this, he had written to Eisig, whose diligent and careful reports had reached him on his trip, »You are gradually becoming the true director of the Institute.« That was not meant literally, but the ironic tone had disappeared: Dohrn knew what Eisig meant to him. He was himself, even though he sometimes scolded himself, an »orderly«, conscientious man, indefatigable in matters of taking notes, preserving things, cataloguing, registering, but it was almost always too much for him: »I must see to it that I can stretch the rubberband of my life as far as possible«, he had written in that letter to Eisig. But in times of depression, this far-stretched rubberband was always on the verge of breaking. Then, seeing Eisig would calm him – the correctness of the office gave him support.

With his inborn politeness which, avoiding subordination, kept him at a distance, Eisig was bound to the later colleagues of the Institute with ever-renewed comradeship. He became the athlete and wanderer of the circle. He rarely had time for his own productive work – he was for Dohrn less a stimulation in their common work than a supervisor of methods. In 1875 he took over the directorship of the laboratories and became indispensable in the inner workings of the Station. He was helpful and loyal, not quite without some sensitivities, but always ready to recognize the value of work, and finally became the true jack-of-all-trades for the running of the Institute. Later in his life, when younger people arrived, he was almost a historic figure in the inventory. He outlived Dohrn by ten years and remained grateful to him because Dohrn, on his part, had continued to worry about Eisig's taking over too much work.

More brilliant than this modest man were the two guests who arrived in Naples at the end of May, 1873. Their work during the following months brought the Station a fame which its founder did not expect to achieve in the early years, but which he later proudly acknowledged: Hans von Marées and Adolf Hildebrand. Through his friendship with the young sculptor, there had already in his Jena years been some contact with the problematic painter. Dohrn, in January 1873, after his sudden trip to Berlin, traveled on his way home by way of Dresden in order to investigate in Saxony the possibility of a work table; there he met Marées and his patron, Conrad Fiedler. His report on the state of affairs fascinated the two men: they heard how the house was growing, and how it was to be a center of intellectual endeavors. Dohrn's lively words awakened in Marées a yearning for Italy, which had until then been dormant.

Dohrn described not only the aquarium and the work rooms, but a common room which was supposed to be the framework for a restful social life and for music. Dohrn described it as opening up to a loggia which faced the sea. The location and the meaning had a claim to beauty, to solemn but cheerful dignity. Was that not a task? Marées suggested that one could perhaps paint frescoes and Dohrn answered, of course one could do that.[23] They talked warmly about this thought. Obviously it was clear to Dohrn that he could not now invest much money in artistic decoration, despite the self-confidence he had acquired through his success in Berlin. He should not worry about this, said Marées. The friendship with Fiedler would give him the support for such an enterprise, and Hildebrand would join them – that would be great.

From the correspondence between Dohrn and Fiedler, it becomes clear that Marées' friend looked at the plan with some worry; Marées' latest paintings had more disturbed than confirmed his expectations. Would Marées' tension be sufficient for such a great enterprise or would the work remain in the stage of improvisation and experiment? »He was rich in beginnings, very rich, but there are relatively few finished works by him. Perhaps – and it is a big perhaps – he will succeed in this case.« The quality of Marées of course remains unquestioned, but Fiedler made it clear to Dohrn that »for the execution of frescoes you will not find in him a very apt person.« Fiedler did not wish to say »No«, but Dohrn should know what was at stake. Dohrn risked it, not so much out of faith in the artist but rather because of the profound eagerness of Marées. Dohrn himself did not have an immediate or an intellectually developed relationship to art. He would be happy to have Hildebrand next to Grant at his side.

The development of the frescoes cannot be told here: their motifs and their great size only came into being in Naples. The first plans, to combine them with allegorical sculptures with painted figures and decorations, was dropped. Dohrn accepted plans which demanded structural change. Where up to now great bow windows had faced the glass-covered court, a long wall was constructed – it now contains the beautiful picture of rowing fishermen. Access to the room went only through the loggia. Where today the door opens from the corridor to the library there was a niche with a marble fountain by Hildebrand (lack of space soon forced Dohrn to turn the festive room into a library). Hildebrand also created the mighty busts of Charles Darwin and Karl Ernst von Baer. When Dohrn placed these men next to each other to honor both, he not only expressed his personal gratitude, which the participation of both men in his work demanded, but also his intention that work in this house should not be dogmatic.

The amount of work by the two artists in these months was enormous. Marées created his own technique of fresco painting and thus freed himself; Hildebrand, too, exchanged chisel and mallet for the brush and started to paint the ornamental pieces. The self-assurance of the artist grew; his letters are cheerful and happy, he enjoyed the circle of friends and nature. They rowed, swam, and hiked. In the evening they sat at Posilippo, in the tavern at the Regina Giovanna, and discussed or silently enjoyed the tiredness of the day with a glass of wine. If they could only help Dohrn! He was, in his liveliness, always a fresh and sometimes a daring companion. There is a story characteristic of his physical constitution and of his will: together with Grant, Hildebrand, and Eisig he arranged a swimming competition from Posilippo to Castel dell'Ovo; a sailing boat was to accompany them. All the others gave up after a third or halfway; he alone finished the long stretch in one-and-a-half hours. That was a great tale. But next to these playful hours there was also sorrow felt by the sensitive friend. Marées would have liked to help by emphasizing that Fiedler would help. Fiedler may have been surprised when he received a letter from the painter saying that he should please lend Dohrn 20 000 francs for the station for a few years. This odd man, who made such grand use of his wealth, had never had any contact with biology. But his sense for the unconventional, for daring enterprise, recognized an elementary force in Dohrn. The loan arrived, and in the finances of that early period it appeared as a lovely arabesque, which a few years later disappeared from the books.

The marvelous monument of this year of creative friendship is the eastern wall of the fresco room, a group picture of the men who are united at the table of the Pergola: Marées and Hildebrand, Grant in a wide-brimmed hat, scrutinizing but self-assured; Kleinenberg standing stiff and tense; Anton Dohrn, separated from the group, leaning on the edge of the table in a somewhat tired pose. There is perhaps in all of German art no group painting comparable to this, in its easygoing freedom and musicality of composition, the keen individualization and the harmonious unity of painting. Dohrn's head is modeled softly, the expression of the lowered glance is somewhat lost, not in dreams, but in a searching intellectual activity which does not completely allow him this hour of rest. The loose position of his body slackens in tiredness – but one feels that it may become tense in demanding and commanding.

The blissful human atmosphere of the Naples beginning was derived, as in Messina, from the Baranowskis. »Whatever they can do, they do it for my sake, so that I really live in Abraham's bosom,« said Dohrn (December 16, 1871, to Kleinenberg). And a few weeks later, also to Kleinenberg, words about their two daughters: »This makes me very happy. Since I own their entire confidence, they tell me all kinds of girlish secrets. Yet they are so kind-hearted that it is never a burden to be with them.« Katharina von Baranowska had not been able to follow her husband to his new Moscow enterprise in 1871, since doctors ordered her to stay in the south. She had heart trouble, and her older sister Olga von Timler took care of her. The women were happy that the German friend brought life, people, and stimulation into the house. They all doted on him, on his freedom from all pettiness, on the idealism of his endeavors. From time to time he appeared to himself, and called himself, a »German bear« in response to the demanding charm of Madame Katharina. Yet he enjoyed the intellectual charm and he felt, in her patriotic passions, a dominating strength which impressed him although he lacked immediate access to it. His need for communication could be sure to find a warm echo in this circle. With Olga, who was somewhat stern, there was some friction and arguments; but life in the Palazzo Torlonia gained daily an almost home-like intimacy. On the first floor lived the four women, on the top floor the few men: Grant and Dohrn, Kelterborn and Lankester, who would have been happy to be drawn into the circle, slightly wondering about his heart, which was attracted to the older girl. Most meals were taken together, the household of the upper floor was lightly controlled by the women's eyes. It had become a pleasant habit for the friends to educate, in an unpedantic way, the daughters, who were also joined by Marie's twin, the delicate Casimir. The use of languages was rich enough, the family spoke Russian and Polish, they had to learn Italian for their daily needs. Anton's correspondence with Katharina was in French, the conversations with the girls were in English. The education of the sisters had been somewhat neglected as a consequence of the early emigration of their parents, who did not want to be separated from them. The son had been sent to a private school. Marie and Casimir were born in 1856, Helene in 1858. This very lack of education, of the conventions of the social class to which the Baranowskis belonged, exerted a special charm, and Dohrn and Grant felt like educators forming these open minds. On long walks the German opened up the realm of Goethe, the Englishman that of Shakespeare. That was not »school«, but a happy game in which these half-childish minds grew, and the two men, who had already worked and fought and were showing the touch of the years, felt moved by the aura of youth.

There were some rumors about them in the colony of foreigners, which reached and disturbed Madame Katharina. Dohrn wrote her at that time (in the summer of 1872) that this was a grateful friendship which united him with the children, and that in his life's plan there was no idea of marriage – he was thinking only of his task and of science. That was honest, but it was also a self-deception as to the strength of his feelings, which grew and changed. Grant saw it more clearly, and he also saw how the older sister's admiration for her intellectual mentor, the fighting hero, changed into affection, love, passion. He played the role of mediator. At the end of 1872 he wrote to Marie: »Dohrn is a man without pettiness, he lives for great goals . . . he loves you in a way he has never loved anybody and in which few women have ever been loved.« She should not worry about his depressions – they were passing. After Dohrn returned from his tense trip to Berlin, they talked to each other. That was in February, 1873.

Katharina von Baranowska was deeply affected by this news – this was not her intention. She herself was too strongly tied to Dohrn to wish or seek a break with him, but her daughter was not to bind her fate of 17 to the older man, not now, not to the uncertain daring, which to care for in a friendly way was very different from perhaps becoming its victim. And did Dohrn not belong to a people which had caused the Poles so much pain? This argument confused the girl greatly, as she wrote in her old age. The right thing was probably to leave Naples, to join her father in Moscow and relieve his loneliness. But the resistance of the parents also awakened the defiance of the girl, the will for happiness which she would choose herself rather than being ordered according to the goals and considerations of her parents.

Dohrn had not expected this situation. He believed that the development had taken an almost natural direction. In Hökendorf, too, his intention was not greeted happily – the distrust was not directed against the girl. But his father had seemingly forgotten his emotion of the past year in Lucca, when he had been almost surprised by the loving care of this son. Now he thought only with uncomfortable worry about the stories from Naples.

10 NEW NEEDS AND NEW HELP

The idea of renting out work tables showed a limited way out of the maze of financial worries, and not only for the moment. This way into the future could be built on a broader and more secure basis. It finally became decisive for the particular character of the growing Institute to constitute a trans-national meeting point for biological scientists, to found a continuity, either for the continuous work by some scientists (how many of them came later with regularity!) or for deepening certain given problems. Of course, the broadening, renewal, and security of the rentals would cause much work, travel, and letters; it would always be important to look for points of contact, to play with connections, to overcome disappointments. But one thing was certain: more important than all canvassing was the achievement of the Station. The satisfaction of the research visitors was decisive for the good name of the Station. This was an uncertain factor. One had to count on unknown degrees of stupidity, arrogance, lack of cooperation, but also on the possible danger of one's own lack of success, of technical imperfections, of the inexperience of the lab assistants. The latter could not just be found anywhere, in the best case they had to be educated patiently and skillfully in Naples.

The positive answer from Cambridge was followed by Falk's readiness. If the formal contract concerning the two Prussian tables could follow only in October, 1873, the word of the Minister was a good lever for the Roman discussions as well as the other efforts within Germany. In Baden, Dohrn had an easy success – he did not lack personal connections; in Karlsruhe Dohrn's brother-in-law Gustav Wendt had been called by Secretary Jolly to become director of school administration; the Grand-Duke Friedrich I was himself interested in the plan. A request to the Bavarian King was well prepared by the position of Siebold and Liebig – in Munich they made a contract for five years. However, the attempts in Saxony and Württemberg did not succeed – the physiologist Ludwig in Leipzig and the zoologist Leydig in Tübingen did not agree with the enterprise. Dohrn may have been consoled when a few years later, both men changed their opinions completely. A certain compensation was the fact that, following the idea of the zoologist O. Schmidt, the newly founded University of Strasburg was endowed with a table in Naples. With this the German program was taken care of.

Dohrn had placed much hope in the responses of Austria and the Scandinavian States. A few years ago, the Secretary of Education in Vienna, Stremayr, had treated Carl Vogt's plans for Trieste with good will; now Dohrn encountered the evasiveness of the new professor, Claus, who wrote him that the Trieste station would probably be founded and then there could be a mutual type of exchange of work tables. This was a suggestion with which Dohrn could not collaborate because of the structure of his own finances; he answered it with some distrust. At this point the great Viennese surgeon Billroth, supported by the other Austrian universities, moved the Crown Prince Rudolf; thus Claus' countergame was defeated. Sweden and Norway were disappointing. Already their first answer gave a general declaration of sympathy, but practically said »No«. Upon the inducement of the Swedish Queen who, two years later, visited the Naples Station, Dohrn renewed his efforts. But Scandinavia, as an effect of the Naples results, decided to start its own Station, which always kept its distance from Dohrn's Institute. Some scientists regretted this and held Sven Lovén responsible. Dutch support was easier; not immediately, but at a meeting with the Leyden zoologist Selenka, during Dohrn's vacation in St. Moritz in July, the firm promise was given to procure everything necessary. What would happen in Switzerland? Carl Vogt answered that it was impossible, that there was no administrative post which could take this on formally. Later he himself would be resourceful enough to construct such a position after he could show others that it was lacking. Vogt's fresh letters also gave, in other ways, advice and encouragement. Dohrn had asked his Geneva friend for his far-reaching knowledge of personnel – but how could he believe that France, barely two years after its peace with Germany, could return to an international scientific cooperation which had been created on Prussia's initiative? There was no lack of scientists in France who considered Dohrn's plan correct and necessary. The following years led there also to the foundation of some Stations. But these, in as far as important work was done, remained within the framework of universities.

Since 1871 Dohrn had corresponded diligently with the American zoologist Louis Agassiz and explained his plans to him. He had received a promise of help. The work tables played no role at that time, but the books did! In 1873, it was a question of the tables – who should rent them? Agassiz thought that for this there was no governmental department, nor a corporation. Here, too, at a much later date, when the need was no longer urgent, a place would be found which would make it possible to send American scientists to Naples for regular work. Agassiz' own attempt, with the help of private money, to create a parallel Institute on Penikese Island was greeted positively by Dohrn – he had, after all, announced that such stations would soon appear all over the world, and had wanted to become the strategist of this scientific campaign on the ocean coasts. Penikese, however, could not be continued.

The efforts of his summer travels to win over Oxford in addition to Cambridge were for the time being unsuccessful. The fact that E. Ray Lankester was an assistant professor there seemed to offer possibilities, and he was in fact a willing spokesman. But

Oxford still resisted; it did not want to forego its exclusively humanistic-classical and church tradition. Let Cambridge do this, where realistic tendencies had found an entrance. Nevertheless, a visit to Bradford where the *British Association* was meeting brought the desired result. That learned corporation completed the contract begun during the preceding year. Negotiations with Russia were still pending. Karl Ernst von Baer had undertaken them, but it became clear that a written correspondence alone was not enough to further matters at the ministry.

As a result of letters and trips, ten work tables could be counted on in 1873, each with 500 Thaler. These sums, whose terms of payment were not yet firm, could nevertheless be counted as a basis for loans. This was a positive step. But the growing organizational planning threw over the present groundplan completely. It had taken only six work tables into account. But the walls were already built, and now they would have to be rebuilt. In the music room the wall for the great frescoes was already finished. But had not this room, which was to be next to Dohrn's apartment, lost its purpose of serving for celebration and recreation? They calculated and measured – there was no doubt that there was no longer room for the intended apartment if the work spaces were to be sufficiently large and numerous. They had to be constructed, too; for each of them a small table aquarium with fresh and ocean water was provided, each with running water.

It was lucky that Dohrn had deviated from Lloyd's plans at one point, and had stuck to the plan of a high reservoir which now made the running water supply possible at the work tables, particularly since some of them had to be housed on the third floor. Lloyd had declared this construction unnecessary, with its necessarily stronger substructure. A simple day and night pump would take care of the necessary circulation. It was also sufficient for the aquarium. For the complete provision Dohrn's solution proved to be much more advantageous.

But on the way to this experience much money had to be spent, which had not been calculated in those bold plans of the year 1871;

then, he had claimed to be able to build the Institute with 35 000 Thaler.

All technical objects – pumps, pipes, sheet glass – had to be imported from abroad, because the Italy of that period did not make them, nor did it have the people who could work with them. It was obvious that Lloyd, with his knowledge of the fabrication of these things, looked to England: an Englishman had to mount and start these things. Thus Mr. Digan became an important figure of the early technical history. Of course, he spoke not a word of Italian; there were eternal fights with the Italian construction workers, which were fortunately limited by not understanding what opinion the partner had of them. But Digan was willing, diligent, he found answers; finally, ambition seized him to overcome the difficulties.

And there were always new difficulties. For instance, there was the question of whether the sea water at the beach was clean enough to be pumped into the reservoirs. The beach was still adjacent to the Station, but one could foresee that the planned road would be built. What should be done about it? At that time the unlimited taking in of water would not be so simple. It was important to make arrangements for this future problem in advance. First, they bought a long and expensive hose with an inner metal spiral which reached places beyond the polluted zone. Then arose the task of waterproofing the stock reservoirs and the aquarium basins. Should one use cement? That was probably the best material, but there was no cement in Italy at that time, and that meant transportation, customs, delay. It was simpler to caulk, as was the local custom. Profumo recommended using *battuto,* a mixture of calcium and pumice, called Lapilli. It worked easily and quickly, and in July the reservoirs and basins were ready to be filled. But after a day and a half the water leaked through and the house was swimming from the dripping water – the time, money, and effort had been in vain. Cement must be ordered. And when the sheet glass arrived, exactly measured and cut, much of it was broken. How annoying! If only a little corner had splintered off, the whole sheet was useless. They saved part of the expense by using cast iron pillars, but that, too, meant much effort.

In the midst of such worries and vain efforts, great fatigue set in. To preserve his health, Dohrn had to leave the south before the hot months. He also had questions about his health. Now that he was busy, the shadow of depression was particularly bad, and it grew into hypochondriac fears. Leyden in Strasburg, whom he consulted, relieved him of these fears. He sent him to St. Moritz, and forbade him to work or to get upset. This was very good advice, but useless for the man who had to read Eisig's reports about the state of the Naples affairs and who had to decide and command. Above all, he worried how his relationship to the Baranowski family would develop. St. Moritz, where Kleinenberg had accompanied him, did not cure him. Then the doctor sent him to Brighton, to the English beach. There, in the company of Grant, Dohrn should not be alone. Brighton did him a lot of good, and a strange encounter gave him back his courage. He met Huxley, the beloved friend, who, as he wrote, being a »physically and mentally powerful man«, was in convalescence from a severe depression. What Huxley told him about his own feeling of intellectual and moral weakness, of self-reproaches which hurt him, was like a report on his own situation.

At the end of September he went to Stettin, where he had a bitter argument with his father: »Do you know who you are? You are a King Lear.« Carl August gave up his opposition to the planned marriage. Dohrn did not stay in Berlin; in Leipzig Conrad Fiedler promised definitely to take over the expense for the frescoes, which was a relief; and in Jena he visited his friend Abbe – this human friendship had reached the point where it was to become fruitful for science.

IV THE STRUGGLE FOR THE PROJECT

1 THE STATION IS OPENED

Scientists were already arriving when Anton Dohrn returned to Naples in October 1873. The first scientist for the Prussian table, which was now occupied, was Wilhelm Waldeyer, who later, as anatomist of the University of Berlin and permanent Secretary of the Academy, preserved a friendly interest in the growing Station. Now, he got to know it in its embryonic phase, but felt its potential for development. Soon afterwards two young Englishmen followed, Dew Smith and Frank Balfour. With a certain joy, but not quite without envy, Dohrn saw these men go to work in the circumstances which he had made possible. He himself would have liked to sit at the table and the microscope. He had just received a letter from his old friend, F. A. Lange, who spoke about the necessity for experiments in zoology, and agreed to the task which his former disciple had created in Naples. But Dohrn hat now become a businessman, diplomat, politician, and financier, with an untiring stubbornness. Basically, only the tools were to be created with which he himself, among others, should contribute to biological knowledge, finding the foundation and the proof of his yet uncertain hypotheses. But that could not be done now. He did see a field of work – the last investigations he had carried out at home were concerned with the perch and the pike. A work by Gegenbaur had taken up the shark as an exemplary object of research; and the selachians began to occupy Dohrn in those years, although from an entirely different point of view. Precisely toward this group the first studies of Balfour were directed; he applied the morphological view to embryological research.

It also happened later that Dohrn's fields of research, which he had so to speak reserved for himself – provided time and mind

would finally be free for them – were taken over by others, Kovalevsky and Semper. And although he recognized their results, there appeared in his letters a nervous malaise that he would be too late with his own production. This was different in the case of the young Englishman, who was only 23 years old. In this case a warm, human friendship developed very quickly. Francis Maitland Balfour, nephew of the next British Prime Minister, Marquess of Salisbury, brother of the future statesmen, Arthur and Gerald Balfour, became an important person for the early history of the Zoological Station in several respects. His studies on shark embryos were the first results of the new research facility, which had a visible effect abroad. Balfour came regularly for some months to occupy the Cambridge table. Dohrn must have felt good when Charles Darwin wrote, in 1875, how much England owed already to the Station, thanks to the works of Ray Lankester and Balfour. Darwin and Huxley saw in Balfour the guarantor of the scientific future of their country. There is something moving in a letter written by the old Darwin shortly before his death in 1882, in which he spoke to Dohrn about a dangerous sickness of his young friend (Balfour) and ends by saying: »Good Heavens, what a loss he would be to science and to his many loving friends.« Balfour recovered. A few months later, however, in July 1882, he met an early death when, during a climb on Mont Blanc, he tried to save a guide after an accident.

Balfour had, with his youthful energy, made the worries of the Station his own. He came to know the Institute in its state of incompleteness but he realized the importance of the foundation. He sympathized, in a tactful way, with the troubles in which Dohrn found himself in order to provide the apparatus. About 1000 sterling were needed. Balfour quickly succeeded in forming a small committee of English scientists which the old benefactors such as Lyell and Lubbock joined – in a few weeks the sum was raised in full (even Arthur Balfour contributed). The correspondence between Dohrn and Balfour has the confidential tone of a close friendship. After the publication of Balfour's *Comparative Embryology*, a new chair of morphology was created for him in Cambridge.

In their letters, they did not discuss science, but rather many different things in a friendly way. Balfour, while realizing the importance of Dohrn's theses, did not follow them. When, during Dohrn's absence, financial troubles occurred, everyone could turn to him. He always helped out until the matter was clarified. When the official transfer of money was delayed, he paid the sums out of his own pocket in the interim. He also contributed to the support of Grant, as did Dew Smith. His eagerness to help sometimes went beyond the goal, because he was not a skilful politician. When, for example, he wished to help his friend during his German difficulties by publishing them in the English periodical, *Nature*, he made matters worse. At the moving news of Balfour's early death, Dohrn wrote to Eisig: »an immense loss . . .! The best of all! What will Kleinenberg say?? For us, his death has very sad personal consequences. For me personally, the only person who was a decent scientific opponent is lost.« He named the second steamer he was able to put in the service of the Station »Frank Balfour,« which kept the name of the suddenly deceased connected with the Station. This seemed to him an objective and human duty.

Two months after the work had begun in the laboratories, the Aquarium, which had caused a lot of trouble, was able to open its doors. Dohrn was by nature certainly not shy – in his letters he often used the word »bold« for his wishes and his persuasion – but he had only a slight idea of the elements of propaganda in the public sphere. A few years later he learned to master it well. But the beginning was without fanfare. Perhaps he did not want to remind the Neapolitan population that barely a year had passed since they had demanded the destruction of the building. To be sure, he no longer needed to worry about hostility. A friendly idea had grown up among the population that, since fish were always delivered to the house, it was something like a »fish hospital.« But in the beginning Dohrn did not advertise, he rather waited to see who would come. On January 12, 1874 he wrote to the Vienna zoologist Claus: »The Aquarium has been opened. The participation of the public is mediocre: there are few foreigners here – cholera, money crises, etc., have trimmed their ranks. Now the English elections to parliament have begun and all the Englishmen are gone.« The theme of cholera recurs in the letters of the 1870's and 1880's several times, but in the calculation it had not been mentioned. Twenty-six people appeared on the first day, a hundred on the second, sixty to seventy on the third, and then the number decreased again. The disappointment of Dohrn, who was so inexperienced, was great. Predictions based on comparison with the successes of the Hamburg and Berlin enterprises had been premature. Patience was needed until it became known in the world that something incomparable, and of a strange beauty, had happened here.

Although he had omitted to announce their new property to the Neapolitans, he nevertheless did not forego telling the world about it. The advertising he had disdained was after all indispensable. And Dohrn understood how important it was to put the Station quickly into Baedeker.* That was successful in 1874. It was a financial deposit for the future, although the daily change which went into the money box was more than nothing. This task remained.

But it was also the fate of this man, so possessed by his goal, that as soon as the Aquarium was finished he was planning an enlargement. The first challenges were overcome, but he was ready to dare to undertake new ones. He was warned by friends, ridiculed by opponents, surrounded by the suspicion that the whole would amount to nothing but a doubtful enterprise which lacked the right intellectual and financial basis. He himself, standing in the center of the action, did not feel it was so much a romantic adventure as a logical necessity.

---

* Baedeker is the well-known detailed travel guide, through many countries and in many languages.

2 »COMMIS VOYAGEUR«*

In a letter of February, 1875, which recounts several insect stories and which contains »Greetings to the Noble Knight of the Station of Suffering,« Carl August Dohrn told his son: »By now it seems to me that you ought to be tired of acting as Commis Voyageur of Vincamparette père – I mean the Station.« His son was acquainted with that tone and did not like it very much, since he was never without worries. Later, looking back, he used the words himself. But in his irony some satisfaction was mixed: his extensive traveling, which others mocked, and which was often painful for him, as well as making things difficult at the Station when life was not yet going smoothly, was nevertheless both worthwhile and indispensable.

Carl Vogt had been right: real capitalists are hard to find for such things. Vogt himself had succeeded in getting 1000 Thaler for Naples from a Frankfurt banker who did not know Dohrn. In return, the Station was to send preserved animals to the Senckenberg Museum** without mentioning his name. The Leipzig professor, Czermak, had helped with 5000 Thaler, a loan which had a time limit; his widow added another 2000 Thaler – that was friendly idealism. There was also Conrad Fiedler's loan, which was independent of the frescoes. To the distrustful opponents who wondered whether, with so little finances, such a decoration was not an irresponsible luxury, Dohrn could always answer that not a penny donated for science had been used for this purpose. It was lucky that the Italian architect Profumo not only did good work, but was also patient about payment. This decent man was up to this unusual task and felt that the fame of his accomplishment would become useful to him later. He remained a loyal counselor to Dohrn. Also, the dowry of 30000 lire which Mr. von Baranowski gave his daughter went, with his consent, into the Station. His opposition

* Traveling salesman.

** Senckenberg Museum, a famous natural museum in Frankfurt.

to the marriage had not diminished his respect for the energy and the enterprise of Dohrn.

In the meantime, Dohrn's attempt had found two new friends. One of them, the new Ambassador to Rome, Robert von Keudell, was to become of major importance for the fate of the Station; the other was Field Marshal Moltke.* He visited Italy in the spring of 1874 and asked Dohrn to show him in detail the Aquarium and the Station. He was obviously greatly impressed by the visit; he also heard some things about the tense situation and expressed the idea that the Station, similar to the Archeological Institute in Rome, should be taken over by the German Reich – he would be able to persuade the old Emperor. This idea was then considered in detail by Keudell and Dohrn. How would it be possible to realize it without robbing the directorship of its unbureaucratic element and the learned participation of its international character? This plan was presented to Rudolf von Delbrück, the German Secretary of Education. Dohrn could inform him of the good opinion of the Prussian Administration of Education and also of the friendly mood of Congress, but he met with the absolute resistance of the Secretary. He would never lend his hand for *that*. This was no solution. It is not quite clear whether Dohrn and Keudell realized at that time that such a change would not have fitted well into the basic Neapolitan contract. Delbrück's agreement would have saved Dohrn some of the difficulties which still awaited him in the future, but it probably would have destroyed the unrestricted development of the entire enterprise, which was only possible because the fight for existence and scientific development out of new sources had to be undertaken again and again.

Delbrück decided on a different approach. During a congressional meeting in November, 1874, he asked to see representative Heinrich Dohrn: Heinrich should tell his brother that the idea of

* Helmuth, Count Moltke (1800–1891). Prussian General Field Marshal since 1871. Educated in Denmark. After 1864 in Prussian service. Leader of modern military operations.

an Institute of the Reich was out, but that he was willing to grant another sum of 30 000 Reichsmark. And he added, »It is terrible – Keudell does everything your brother wants.« This groan was not quite genuine. For it was precisely Dohrn's tenacity which had awakened Delbrück's human sympathy. Dohrn had fought first, in more than one way, on the volcanic soil of Naples, and then against academies and departments. When Delbrück, a little later, stayed in Naples, it was Dohrn's task to make him acquainted not only with the Station and its work, but also with Naples. The personal relationship which thus became closer promised the best for the future.

Heinrich Dohrn had been elected in 1873 to the Prussian House of Representatives and soon afterwards to the German Congress. The office of parliamentarian which, for his father in 1859, had not been more than a »noble office,« took on for him the nature of an inner passion (although he covered it up with skeptical composure). As a young man, in 1867 he had taken part in the foundation of the national liberal party, and as a result of his public interests he became an honorary member of the Stettin Magistrature. He followed his father in his concern for entomological interests, and later as president of the Entomological Society. He also dedicated himself to commercial enterprises, being more speculative than Carl August but, as it became clear later, he lacked sufficient insight and patience for this. His local work was very meritorious, as was his dedication to educational problems and the broad education of the lower classes. Later, he went beyond entomology and turned to antique sculpture. The bronze reproductions from Antiquity in the Stettin Museum are essentially due to his organization. The question of whether his eagerness for broader political work corresponded to his talents remains unanswered.

Heinrich Dohrn's main features were a sarcastic self-centeredness and a shy dogmatism. Toward his younger brother he wavered between a superior stance of knowledge and great admiration. He had a great and clear view of Anton's task, but did not like to present it to their father for the sake of peace at home. He lived, as an unmarried man, at the home of his parents. Heinrich's entrance into the parliamentary communities seemed at first to be a great advantage. Just in the years when the relationship of the Station to the Reich needed some clarification, it was good to gain a light, unofficial connection with the representatives, particularly with the members of the financial committee. The »Commis Voyageur« learned of new addresses to which he could turn. Later, Dohrn experienced the bad side of the fact that one had a brother in parliament: this happened when Bismarck's angry glance was resting on him.

The good news which, in November 1874, Heinrich had to communicate to his brother about Delbrück's willingness reached Anton when he had returned exhausted from several summer trips and realized that the next summer would demand more trips. Both trips were real round-trips, serving essentially the purpose of gaining and securing work tables.

The most important trip was to Russia. On June 3, 1874, the wedding of Dohrn and Marie Baranowska had taken place in Warsaw, »first orthodox, then Lutheran«. The wedding was under the shadow of the serious illness of his sister-in-law, Helene. The next stop on his trip was St. Petersburg. A year earlier, the request for participation in the Station had been sent. Karl Ernst von Baer offered his intervention at the Russian Academy. But he felt, having had some lack of success himself when he had tried to get help for Dorpat, that Dohrn should himself support his own case: »Your own initiative could have more success. If you knew Russia well, you would know that a foreigner could be more successful in his request than a born Russian. The government wants to be praised abroad. This is the basis for the Sinai Bibles and similar events.« And indeed, Dohrn made some progress. The Director of the Education Department was of good will, the Secretary of Finance was immediately won over. A recommendation from Baranowski opened the door; the German Ambassador, Prince Reuss, offered his help. In the three weeks of visits and dinners Dohrn's ideas of Russian strength developed into a sense of what he would have to deal with from now on. Trips in Germany produced a table for

Saxony through a petition to the King, and the alternate participation of Hamburg and Hesse. In Stuttgart, a later success was initiated. Siebold was able to report that the Bavarian scientists spoke of Naples with great satisfaction.

If one reconstructs from Dohrn's notes the summer trip of 1875, one finds the following succession of places: Vienna, Budapest, Galicia, Breslau, Stettin, Danzig, Königsberg, Berlin, Hamburg, Bonn, Cambridge, London, Stockholm, Stettin, Leipzig, St. Petersburg, Moscow, Berlin, Hamburg, Bremen, Oldenburg, Stettin, Naples – almost everywhere he had negotiations, in some of them talks, sometimes successful, sometimes disappointing. Dohrn found out that his work was beginning to be known, that persons who earlier had refused now agreed with the matter, but that gossip, disfavor, and misunderstandings had also come about. The trip had served its purpose of winning new tables. Things were hopeful in Hungary; Scandinavia was reserved. The financial support of private persons, which he had expected for a steamer, always remained below expectations despite optimistic assurances. But at the Academy he had a victory: it donated the necessary steamer to the Station.

Some people suspected or laughed at his constant traveling existence. It was also described as a somewhat expensive luxury which did not always produce results. But in 1872–73 he had tried out the effect of personal intervention in matters of »dangerous weapons,« and he knew that more than the many memoirs he might compose (»per mitrailleuse« *), the spoken word clarified, enthusiasm was evoked, and unafraid frankness disarmed his opponents. He learned that one had to accept useless stops, hours of waiting in Ministries, patient listening to unknowledgeable advice, if one needed the adviser for other things. Of course, he was at times sick and tired of this »Commis Voyageur« life, as Carl August had guessed. He complained in 1877 in a report to Eisig that »one cannot tolerate intellectual veal, particularly re-heated, without

* Mitrailleuse is a machine gun.

getting terribly bored.« But a few years later, in 1882, Eisig received a note of gay, ironical triumph from Frankfurt, where things seemed to advance well: »I have of course lived through a great amount of dinners and suppers, but those are the battle fields on which victories are won, which help us to continue.«

The correspondence with his old friends Stahr-Lewald had become rarer in the frantic years of the Naples beginning. There was no calm for self-reflection, for which the couple had provided an attentive, benevolent, and critical public. He also had outgrown their measure of worries as well as success. His decision to choose Naples had scared rather than calmed them. They did not trust the strength of his persistence. But a letter of April 30, 1876, half a year before Stahr's death, takes up the old tone again. In it, Dohrn characterizes his uncertain and paradoxical inner situation: »Certainly the enterprise eats up my best strength and interest. But it is so grand, and it develops in such gigantic proportions, that despite this decay created by business, I often feel great joy, which is the best reward and cause of creativity. It is something unique, something free and independent, created entirely without example, merely by my own thought, and it carries creative traces of every phase of my existence. It has produced what any factory owner produces who starts a large business for European demands – and it was created under such demanding circumstances. You see under what difficult conditions it was created. I am lying to myself that it was really the goal of my life to create the large laboratory. That is not true: I wanted something else. Will I reach it? I wanted to create something intellectual, to live in it, and finish my life within my inner education. But nowadays this is more difficult than in former years. The racing wild life of the present carries us with it, God knows where, and it includes even those who resist. And one has enough to do to protect oneself and one's family, in seeing that it does not race over us.«

3 THE ORGANIZATION AND GROWTH OF THE STATION

When Anton Dohrn wrote in the spring of 1876 of the »enormous proportions« of the Station's development, the work tables had already been occupied more than sixty times in barely two and a half years. The stay of the individual scientists differed between a few weeks and half a year. In the summaries which were enclosed in the *Mittheilungen*,[24] edited since 1879, this is noted down accurately, including the publications which came out of the Naples studies. This is not unimportant for the expert. Of equal importance for the scientific usage of the Institute was a question by Johannes Reinke, who at that time was teaching in Göttingen, whether there would be room and a chance to work for botanists. Dohrn answered this very positively, and Reinke worked in the winter of 1875–76 on algae.[25] In a report to Secretary Falk he wrote that his boldest hopes had been surpassed: »All the research was only possible due to the arrangements of the Zoological Station. These are grand in manner and in every respect outstanding. I am happy to have, through my initiative, won this Institute for botany.« Dohrn's supporters in the department, such as Goeppert, liked to read such judgments. Reinke's visit received fundamental importance, a botanist stopped among the assistants, the marine flora entered the realm of research and later publications. When diving apparatus was put into the service of research, it was meant mainly for botanical needs.

But what was of course more important for the Aquarium, as well as for the work tables, was the acquisition of »sea beasts«, as Dohrn liked to call them. The Station had promised to put these reliably at the disposal of science. That meant a twofold educational task. For one thing, the scientists had to learn that they just could not come suddenly, but rather had to announce in advance which fish or crayfish, which larvae they wanted – and they had to accept the fact that perhaps their wishes could not be fulfilled at all in that particular season. In order to clarify this, Dohrn published an

»open letter« in Siebold's *Zeitschrift für Wissenschaftliche Zoologie* (July 1875), which happily and practically talked of these matters. The lively essay also contained a characteristic analysis: the scientists visit the Aquarium too rarely. They are sitting with microscope and microtome at their work tables and believe that the lower floors serve only a curious public of any description. Dohrn himself had felt that way in his first illusionary financial plans. Much later, he realized that visiting the Aquarium was not such a bad idea after all. For himself, each visit before the glass panes with their background of rich fairy tale forms stimulated exact observation and investigative imagination. One found problems and considered methods of how one might penetrate the true laws of life. To educate the scientists, which in the beginning had its painful or funny episodes, was not easy, because the habits and expectations of the various nations accompanied them to the Institute. But there, too, a tradition arose.

It was more difficult to create a habit among the Neapolitan fishermen. The work depended on them; some sailors had to be taken completely into the service of the Station. For the technical problems one needed a solid group. It was difficult to educate these people and to make them work together. There was no doubt that Dohrn's relationship with the Neapolitans never lacked authority – he had almost too much. They came to him with their fights, wishes, family stories, and jealousies. He later thought that bringing about a dependable order in these matters had sometimes cost him more nervous strength than the painful worries about the money for construction and wages. Of course, this situation was part of the finances which, in his touching early calculation, had not been foreseen. One could not send the fishermen out to catch this or that. With the ocean fish used for trade or food, certain fishing grounds were in the possession of traditional owners. But that was not enough. What one wanted most were other sea beasts: sea urchins, seahorses, crayfish, pelagian animals, mollusks, etc., which were »useless« and had no trade value, and were normally thrown back as unnecessary weight. But they were precisely what was needed in the Station, certainly not all and not every day, but nevertheless fishermen of Mergellina, also those of Procida and Ischia, were taught to offer all these by-products of their catches, and Dohrn was determined always to buy them all. Around the Station something like a colorful, loud fish market arose, a noisy bargaining about things that once had no market value but had now suddenly acquired it, to the amazement of the people. The fishermen employed by the Station gradually became specialists, not only with regard to the treatment of people, but also in the needs of science. Dohrn spoke in his reports with a certain admiration for these men, whom he first attracted and then kept for decades. Charles Grant's empathy for popular customs was in this regard a particular support, and Marie Dohrn was often indispensable, since she knew the Neapolitan dialect best.

It was soon very clear that the boats with which they began – a sailing boat and a few small vessels – were not sufficient for systematic catches. They had to become independent of weather, they must be able to reach the islands, the Gulfs of Salerno and Gaeta, and »dredging« needed to be made easier. A »dredge« is a dragnet used to raise the animals living at the bottom of the sea and the sea bottom itself, with its fauna and flora. A fast steamer, independent of wind and weather, whose power could take care of this painful dredging, seemed indispensable. Since 1873 Dohrn had thought about this, looking around to see whether an old boat (on account of the expense) was to be found somewhere, into which one could build the dredging apparatus. A letter to du Bois-Reymond on July 24, 1873, full of plans, reported that he was looking for such a steamboat – its cost could be about 4000 Thaler and it must be »self-supporting,« since it would be rented out to tourists for one- or two-week trips in the gulf and along the coasts. He also mentioned a plan for a railroad car with an aquarium, to transport marine animals to aquariums and universities! That was a fanfare. Now, in 1875, the problem of the steamboat would be handled with energy; it was the main reason for his visit to London. Werner Siemens had referred him to his younger brother, and Wilhelm

Siemens (at that time he was not yet Sir William) became Dohrn's unselfish counselor. Through his important inventions concerning the production of steel, Wilhelm Siemens had acquired a leading position in the intellectual and industrial life of England. His life, which led him to foreign countries, had been daring. Now he turned his helpful sympathy toward his young countryman. When finally the commission was given to the Thornycroft wharf, the technical advice of Wilhelm Siemens accompanied its execution (the wharf had become famous a little earlier through the construction of torpedo boats). Dohrn himself had been directed to England when he inquired about his plan in Hamburg and Stettin – this is characteristic of the times.

In the meantime, Werner Siemens had become an equally important help. The special expense for the steamboat, which would amount to approximately 20 000 RM, could not be paid out of the small daily income. The few special gifts of that time were eaten up by older obligations. Could one organize a special collection for the purpose, as Balfour had succeeded in asking of his British friends? There was no lack of willingness to found committees, particularly since Helmholtz, du Bois-Reymond, Virchow, Siebold, and Leuckart signed an appeal to the public, which underlined the English enterprise. In Hamburg a donation of a few thousand Marks was offered, but it would take a long time before the friendly approbation Dohrn had met with in 1875 in Breslau and the Rhineland would grow into a sufficiently large account. Perhaps Crown Prince Friedrich Wilhelm could help again, as he had done in 1873? In April, 1875 he had made a surprise visit to the Station and was obviously very satisfied. When Dohrn asked Helmholtz, the latter advised him not to bother the Prince with money problems. He knew that Friedrich Wilhelm was more generous with his benevolence than with his money.

Ultimately, it was Hermann Helmholtz who made the decisive move. When Dohrn visited him in Berlin at the end of July, Helmholtz suggested to a pleasantly surprised Dohrn that the Academy itself should donate the entire steamboat – he did not know of a better use for its money. That sounded good, and Dohrn prepared a petition. It gave a proud survey of the first years: it turned out that instead of four and then ten work tables for foreigners, as they had originally planned, now 24 tables »are not too many«. And after the account on the costs for building and management, he quoted the words which the English biologist Richard Owen had said to him: »You have created a new dynamic in science.«

But in the meeting of November 8, 1875, the petition was first rejected – the main opponent was Theodor Mommsen. This lifted the event above the realm of the incidental: in a certain sense it was a test of power between the mathematical-physical section (du Bois-Reymond) and the historic-philosophical section whose permanent secretary Mommsen had just become in 1874. The little steamboat, which was in the center, became almost a symbol of the period – it became a forepost in the quarrels in which the natural sciences tried to secure rank equal to history and philosophy. The charm of the episode is that the captain of the boat, in his universal way of considering life, was closer to some of his opponents in the Academy than they realized.

A suggestion for a compromise by du Bois-Reymond saved the situation: the Academy should rent a work table for ten years, but pay for it in advance with the sum of 18000 RM and petition the Secretary of Education for a donation of 6000 RM. This was done, although the solution dragged on for a while. When the members opposed to Dohrn (those whom he had found to be quietly hostile as early as 1872–73) protested again, Werner Siemens, in the decisive meeting on December 21, went all out for him and took over the responsibility for a professional procedure. Mommsen mumbled and remained privately hostile, although he no longer emphasized the fact. Years later a friendly relationship of mutual visits between Dohrn and Mommsen developed. In Dohrn's correspondence, Mommsen's *Roman History* is mentioned often – but Dohrn's admiration for the literary achievement does not hide a certain distrust of the subjectivity of Mommsen's judgments.

The much-hoped-for steamboat arrived in Naples in 1877. Dohrn named it »Johannes Müller«, thus honoring the great physiologist as the founder of Marine Biology. The oldest of the fishermen, who now cooperated in the dredging, had been Müller's helper decades ago, when Müller studied *Amphioxus* in Naples. The boat became the cause of worries and joys at the Station – who knew how to handle it professionally? Seven years later, Siemens suggested a thorough overhaul and a new cover of teak wood, in order to save on the eternal repairs and re-paintings. The entire crew of the Station had learned a number of things in those years. Dohrn was no longer alone as at the time when they had to ask Lloyd to send them instructions for the Aquarium plant by mail. Now a broadly knowledgeable technician had entered, the German-Russian von Petersen: architect, machine technician, captain, whatever was needed – a somewhat eccentric man with all kinds of shipwrecks in his life. He liked Dohrn and was ready to throw every unwanted guest out for him. But when Dohrn was out of town there was an air of thunder and lightning around him. He now took care of the machines and the boat.

From the beginning the program of the Station included a first-rate professional library. Its necessity did not require special justification. The visiting scientists could not bring with them books on all their problems. Dohrn's own scientific library formed the core of the growing collection. The eager Huxley saw to it that the Royal Society donated its present and past publications. A visit to Ignaz Döllinger, under Siebold's patronage, resulted in the gift of the scientific writings of the Bavarian Academy. Berlin did the same. Publishers and authors were advised to donate. The catalog of the holdings in early 1874, as supplement to Siebold's journal, already makes an excellent impression. One consideration was that there is a limit to generous donations. The budget held on to a definite sum. In it appeared a special donation of 5000 lire from the Italian Department of Education. But that was not enough, because, naturally, the entire international professional literature had to be acquired. In a daring but successful literary publication,

Dohrn attacked this question: he created his own scientific journal, the quarterly issues of the *Mittheilungen*. This became an object of exchange for receiving periodicals. He also founded the *Zoologischer Jahresbericht*. Its direction was originally taken care of by the Leipzig zoologist Victor Carus, but soon Dohrn had it published entirely by the Station. Here, too, the immediate gain was the voluntary mailing of books and reprints, in which the authors were interested because they were included in the yearly report which was organized according to groups. This was of great advantage for the Station. Within a few years the library became internationally famous. At the time of its origin it was incomparable, and probably remained so, although scientific research institutes have since been created in great quantities. More than once, scientists occupied a work table in Naples, not in order to experiment or observe there, but rather to check their own literary work in the Station's library, which was so easy to use.

The *Jahresbericht* was not supposed (in Dohrn's words) to be an annual judgment, but simply a good overview of work in science. The important gain for the library which resulted had, from a practical point of view, been sufficient cause for the attempt.

Both publications continued to be in need of some financial support, but it was a small sum. They offered a place (for publishing) to the assistants and other scientists working in Naples. In the *Jahresbericht* the connection of the work of the Station with the entirety of zoological problems had to be secured. Dohrn hoped to conquer a leading position in scientific publications. For this he had to secure good collaborators and to omit the traditional narrow-mindedness of scientific reviewers. This goal was achieved more or less completely. There existed two older journals, the *Archiv für Naturgeschichte* und the *Zoological Record*, but the former was published irregularly and allowed its reviews to appear sometimes only after several years. The English publication was more reliable, but thematically limited to taxonomy. As Dohrn wrote to the Prussian Academy on May 9, 1881, the Englishmen admitted that the Naples Report was superior to their own publication. But

Dohrn's attempt to combine the three publications did not work out, so he went his own way – the Italian and Russian governments and two Dutch scientific societies gave some money and the Prussian Academy donated, beginning in 1881, a contribution of 2000 RM for a number of years. The *Jahresbericht* distinguished itself by a quick and reliable survey. The true critical review, in other words, the judgment, was being avoided. One could perhaps guess it from the length allowed to the individual contributions. For scientific research the enterprise was of central importance. When it fell victim to World War I, a vacancy occurred which was never again filled with the same objective.

The third series of publications which followed it was the splendid *Fauna and Flora of the Gulf of Naples and Neighboring Waters*. According to a note this plan went back to the year 1875 – it seems to have found its definitive form in connection with a lecture to the Scientific Society in Moscow. The goal was a large inventory of the fauna and flora and the topography of their environment, very useful for research. Above all, the taxonomy of the marine fauna should be organized. A corresponding, unified, and unequivocal possibility of observation had until then not existed. Thus, it came about that Dohrn's greatest literary enterprise was directed toward – systematics, but in a very different sense and measure than had been the custom up to then. A report to the Academy of December, 1877 contains skeptical words about the results of accidental expeditions »which appear like the prospect of any Encyclopedia still to be written.« Here work aimed at a total observation: »Does not the zoologist have to ask the algologist, the geologist and mineralogist, and all three the hydrographer, chemist, and physicist, about any question of conditions for existence of even any one animal?« If one first knows the manner of existence of animals, speculation will disappear which has put the »stamp of being novelistic on physiology and made it lose credibility«. The *Fauna and Flora* was prepared in such a way that special instructions were given to scientists and assistants. Dohrn himself worked on one of the volumes, on pantopods. This matter, too, needed its own financial foundation; it succeeded surprisingly well. The volumes could be subscribed to, and the price was then a third of the eventual retail price. The friends, who some years earlier had been stimulated to form committees for the steamboat, were now shown the *Fauna and Flora*; princes, academies, and also the recent rich industrialists, were encouraged in the friendliest way. There was no longer any begging; rather, if they agreed, they could feel that they were receiving a present. Of course it was troublesome to start the enterprise, above all to win the crowned heads, now also the Spanish and Portuguese kings. By 1914, 33 volumes had been published; those with illustrations became exemplary.

The complex scientific publications of the Station, which was technically in the hands of the new assistant, Paul Mayer, had grown out of the immediate needs of the library and the systematic survey of the material. Another department also flourished, first only marginally, but when it got into the right hands it gained an importance far beyond the Station. In order to familiarize the fishermen of Mergellina with their delivery, the Station bought their entire catch. Some of the material went to the Aquarium to renew and supplement it, other parts were needed for the work tables of the present scientists. What of the rest? Should one throw it away? That was not a good idea, since one was training the Neapolitans to respect even those things that seemed useless. Thus one started to preserve the animals. Some scientists needed them to continue their studies or wanted them for the purpose of teaching. And was there not also a »market« in the museums of natural history? Dohrn's practical imagination mentioned the treatment of research material as early as the first plans of the Station. He had originally deceived himself about the economic and scientific effects of the Aquarium. But soon the preservation held, quantitatively and principally, a higher rank than had been expected.

Siebold had recommended to Dohrn a sickly student, August Müller, for whom the stay in the south seemed beneficial. Müller started as assistant in 1876. His lung trouble prevented him from doing hard work, but he began, thoughtfully and carefully, to improve the methods of preservation with new chemical experiments. He was given as a helper a young Italian, Torillo. On his father's wish, Dohrn took the boy into the Station when he had finished school at the age of 14. In his intercourse with the fishermen, in the distribution of material, he had proved himself useful and dependable. Now he watched the manipulations of the young German; but this was not enough for him; so he tried the work himself, despite the inherent danger. The young Swiss, Arnold Lang,[26] who had twice occupied the newly founded »table« of the Swiss Society of Naturalists and who entered the full service of the Station for five years in 1879, enjoyed giving systematic private instruction in zoology to the eager young Italian. »Torillo« became a leading figure of the Zoological Station. He was the son of the doorkeeper of the Torlonia palace, where, since 1872, the Baranowskis, Dohrn, Grant, Kleinenberg, and Eisig had been living. They knew the boy, who always sat over books or played or drew corals and other things. But the boy was silent and reserved toward the foreigners who visited the doorkeeper's lodge from time to time. It was simply as a friendly favor that Dohrn fulfilled the wishes of the boy's father, who did not know what profession to choose for the boy. There was always an opportunity for a handyman at the Station. But the young Salvatore Lobianco showed soon enough that intelligence, will, and ambition strove beyond this, and the cheerful nickname was dropped when, after Müller's early death in 1881, the preservation department was totally handed over to the 20-year-old self-taught worker. He now became, with somewhat solemn irony, »Don Salvatore«; the young man stretched himself, gave his department *his* stamp and became Signor Lobianco. But things went further: he started his own scientific publications, and the University of Naples conferred a doctorate on him without bothering with any examinations. The man had grown to be an independent figure of power in the center of this world. He became an original technical genius, inventing and developing methods, but he never did mechanical work in a routine manner. Lobianco's work created for the Neapolitan Zoological Station its worldwide reputation for producing the most beautiful preserved specimens. In the financial accounts, expenses for alcohol grew. The department developed into a real export business, which had to take care of collections and institutes in the five parts of the globe. The question arose of how to respond to foreign requests for study in this department. Dohrn had to think this over carefully. When the time came when new stations would be erected by different states and by learned foundations, he was glad that his prophecies were proved correct. His correspondence shows how lively the participation was, also in matters of advice. But nobody wanted to give away the special experiences gained with so many sacrifices. In 1882 he was already sending instructions by letter to Eisig that preservation was not to be taught to outsiders. Once one had accepted the word »science factory,« there were factory secrets. One wanted to keep the monopoly of this special art and insight as long as possible. Somewhat later, in 1889, the Prussian Academy asked Dohrn to release the Naples methods to the newly planned Hensen plankton expedition. Dohrn did not like this; but he decided to let Lobianco publish a larger presentation of his methods in the *Mittheilungen* of 1890. As long as the connection with the Station remained intact, he himself suggested and took care of the instruction of the newly created art, as will be shown later.

Another side of biological methods received from Naples its decisive support: the microscope technique. In this field the work of Paul Mayer of Lüdenscheid (he had started as an apothecary) achieved a rank corresponding to the position of Lobianco. The contemporary circumstances were favorable and personal relationships gave them a particularly effective character. At the beginning of the 1870's Ernst Abbe in Jena had almost newly founded microscopy by his theoretical statements and his important constructions. Nobody could use it faster and more encompassingly than Dohrn's enterprise, to which the tools for exact research in microbiology were thus offered. But the man in Jena, whose approval in 1870 accompanied Dohrn in his bold enterprise, knew how much the Naples' needs meant to his own beginnings. When he, in 1897 gave the great memorial speech on the 50th anniversary of the existence of the Optical Workshop Carl Zeiss, he gratefully remembered the old friend of the common Jena years, »in order to name at least one«: »From his powerful and creative being some rays entered our home.« The close contact of the two men became professionally fertile. Not only did Abbe give the Station instruments at a very low price or even donated them, he also received from it stimulation, control of the practical aspects, and the critical »co-thinking« of experience. The Station also became the place where foreign scientists first met with Zeiss' products and bought them. They also announced at home the superiority of German work. The Station became a kind of export warehouse; and Dohrn was happy when he could write to Abbe that, for example, Balfour had purchased this apparatus and wanted to have another kind.

The correspondence sometimes went into technical details, and gradually P. Mayer, who at first had been mainly busy taking care of the publications, developed almost pedantically the relationship of preparation and microscope work, thanks to his precision and stubbornness. He wrote the manual which was the leading work on microscope techniques for a long time.

Aside from Eisig, Lobianco and P. Mayer remained Dohrn's collaborators until his death. For many assistants of the later decades the years at the Station were a transition to professorships or museum work. One important person who, beginning in 1881, found his life's profession in Naples, was Wilhelm Giesbrecht from Danzig, who first supported and later took over the literary work of P. Mayer. He was also, through his great technical skill and his easy adaptation (Mayer lacked the latter), a very fortunate element for the atmosphere of the Station. In later years he enjoyed Dohrn's particular human confidence.

Perhaps it was Dohrn's most important task and his greatest achievement to create an »atmosphere«. In Volume 6 of the *Mittheilungen* (1885) he wrote a memorial on the old lady who had sold tickets at the Aquarium since 1874, a Frau von Boutkewitsch, widow of a Russian colonel. She had taken care of her position with brave resignation and a cheerful freedom. She left her small property of 6000 lire to the Station, which had offered her an existence in her old age. Dohrn devoted to the educated and charming woman a friendly characterization: »Legend tells that Rome was created through the getting together of men who had opposed human society and had forced themselves into a situation in which they put all their luck on one card. This was also the fate of the Zoological Station. Starting with its founder, who felt himself, at the age of 30, problematic as a human being and a scientist, down to the men who took care of the Aquarium and the laboratory – they were all more or less shipwrecked men who found themselves together in Naples and who with me as their leader, either started a new life in the most fantastic way or finished an old one in peace ... We all were similar, in some respects, to the »Bremen city musicians« of the fairy tale, who individually did not quite know what to do with themselves, but had an indefinite hope that together, they would work matters out.

In fact, the fates, origins, temperaments were rather different and it was a difficult enterprise to form a unity out of the adventurous group which was thrown together. Kleinenberg did not fit in. But with the others Dohrn's instinct for human beings proved itself true and his leadership won out. This was not a matter of course, for his trips kept him from having immediate influence for months, and periods of paralyzing depression weighed on the Institute. Even the dependable and correct orderliness of Hugo Eisig did not have enough suggestive power to withstand the human problems in an indifferent way. Dohrn demanded extraordinary amounts of work from his co-workers. Not only should they, with their scientific work, give honor to the new Institute, he also delegated to each of them a full measure of administrative work. He gave a task and direction to the scientific investigations, but did not force them into his own or anyone else's zoological dogmatism. The administration consisted of business correspondence, bookkeeping, statistics, and accounting responsibility for their department. Not until 1883, when he himself had reached the end of his strength in administration and was afraid to be totally divorced from his own scientific plans, did he employ a secretary, Hermann Linden. That was a good move: this young man, a complete stranger, had written to him that the Institute probably needed a person trained as a bookkeeper. He wanted to come to him in order to help. He spoke four languages. Only when this man was there did Dohrn realize how much he and his assistants needed him. From that moment on, it was not possible to imagine the Station without Linden. Dohrn could, without paying overly high wages, ask him for much work, because he himself gave the example of total dedication to his work combined with the simplest way of life. It is barely imaginable how much he imposed on himself (after 1874 also on his young wife) concerning technical writing, keeping records of receipts and expenses, protocol notes, etc. When he was traveling, question and answer notes traveled back and forth. Although he encouraged independent decisions, he was, despite great courteousness, of a fundamentally autocratic nature with great will power. His imagination and worries were constantly busy with the whole and the details. Of course there were also crisis situations – he threw people out of the station who did not behave well, and then fought about it valiantly. But he knew how to attract his closer co-workers through social contacts. There were conflicts and difficulties, when it was a question of a wild temper such as the engineer von Petersen and the miserly P. Mayer. The latter received characteristic educational instruction, from a distance, about the treatment of the personnel: »The larger portion of mankind is less than mediocre, and those above average mostly neutralize each other. It is therefore practical to bring together a remarkable person with an unremarkable one – otherwise there will be trouble and the drum receives holes. Perhaps one can improve now and then. Normally one gets a blind person for a lame one, and nothing is improved. Elastic personalities are rare in all classes. Let us keep going as best we can (1882).« A year later, when Mayer complained about Petersen, he wrote: »The art consists in welding together different individuals. I am too conscious of my own weaknesses to condemn others ruthlessly. There is only one quality I cannot tolerate long around me – naked egotism.«

Dohrn himself succeeded in finding some way to Petersen, who as a technician could be used everywhere, not a solid worker, but an »amazing« one. Dohrn could tell him anything, nobody else could. Ultimately even Eisig was no longer on good terms with the Russian. Open conflict was threatening. Dohrn had been away for some time in 1887: »Leave the guidance in this intrigue to me,« he wrote, and continued, »It is lucky that, for some days, my nerves have become better and my old elasticity, which has often saved me, is not lost. Let everyone do his duty as if everyone were responsible for the well-being of the whole. Let everyone forget about his own private preferences, and the matter will work.«

4 SCIENTIFIC WORK AND SCIENTIFIC POLEMICS

In 1875, Anton Dohrn completed the »Genealogical Sketches« which he entitled *The origin of vertebrates and the principle of change in function.* This small article, a little over 80 pages, was dedicated to Karl Ernst von Baer in Dorpat. As a gracious well-wisher of the Naples efforts he accepted the dedication, although the basic theme was not in accordance with his own teleological view.

The author had perceived for quite a while that he should come out with a new publication which would make a statement on the results of his own investigations in Naples. After all, the Station had been created out of the realization that his own research was inadequate. Its systematic pursuit had suffered from the restlessness of the years. Above all, for a considerable time he had worked with the selachians, the sharks; they were particularly important for evolutionary studies, because they are cartilaginous fishes, that is, their skeleton is made of cartilage rather than bone, and this is always the embryonic development of higher forms. The investigation of the articulates had guided Dohrn to the annelids (or worms) as the ancestral group. The vertebrates have some features of their body plan in common with the annelids, above all the segmentation of the body which is apparent in vertebrates in the vertebral column and the nerves emerging from the spinal cord, and even more distinct in embryonic development. Dohrn had developed the theory that this highest phylum of the vertebrates descended from the annelids, and he was determined to provide this decisive contribution to the phylogeny of animals. He was convinced that embryology would provide the evidence for his thesis.

This short article was the prelude to a long series of studies which would be published in the following decades in the *Mittheilungen*. He adhered to this thesis with a certain tenacity even when it gradually became antiquated and the generation of young zoologists and biologists, quite often stimulated by the opportunity for work which he placed at their disposal, went in quite different directions. In this respect he remained the contemporary and student of Haeckel even though in subject matter he became his adversary, and the friendship of his youth turned in part to sarcastic opposition. His aim was also to establish a phylogenetic order, though in a much more cautious way than Haeckel.

In all this, the decisive impulse by Darwin continued to affect both. Dohrn wrote to his sister Anna Wendt after the death of the great Englishman that the Station has the mission to become »the executor of Darwin's testament«. But this claim was contested bitterly in Jena. And, as he wrote to Keudell on May 3, 1880, Dohrn considered this as a »struggle between life and death . . . If I were to try to explain with a metaphor what it is all about, I would say that Darwin has erected a monumental edifice, with powerful foundations and immense walls. But he has left it to the research of the century that followed him to enlarge and to furnish the house. I have taken charge of furnishing the corridors, all staircases, and the largest hall. But my former teachers Haeckel and Gegenbaur wanted to do the same thing at the same time. I reproach them, however, for having built the backstairs instead of the stately front staircases, and their corridors are nothing but blind alleys. They denounce my staircases for being fantastic products which will break down; up to now, the scientific henchmen run after them, with a few talented exceptions, but I have arrived already at the hall, with some secret constructions. And in one or two years it will become apparent who will kick whom downstairs. As for the danger, your Excellency, that you consider me a Don Quixote or a searcher for the perpetuum mobile, I promise you that I will stay in the hall and Haeckel will be thrown down the stairs . . .«

This is certainly not a scientific argument. But the way Dohrn shifts without effort from an architectural metaphor to the image of a brawl characterizes quite well how he feels about this test of strength.

It is not for us to judge the publications on the origin of the vertebrates. While with Gegenbaur and the school of comparative anatomy the emphasis for the assessment of the vertebral system and phylogenetic considerations was on the skeleton, Dohrn started, above all, from the organizational characteristics which must have been present in the ancestors of the vertebrates, that were without skeleton and must have been taken over from them by manifold transformations. In phylogeny, the head must have been subjected to particularly intense transformations. Therefore, Dohrn's studies were directed first to the gills and to the structures originating from the pharynx, such as the thyroid gland, then to the arteries which supply the gills; further to the earliest embryonic stages in which segmented pharyngeal pouches are temporarily formed, like the segmental coelomic parts in the rest of the body, from which the muscle segments originate. Finally and above all he concerned himself for the longest time with the development of the cranial nerves. All these painstaking investigations have the goal of reconstructing an ancestral vertebrate head, which could be connected to a lower form such as the annelids. Furthermore, he investigated the supply of the dogfish fin with musculature and nerves during its embryonic development. The evidence of the sequential origin of these structures is supposed to support the bold hypothesis that the limbs of the ancestral vertebrates are derived from the lateral appendages, the »parapodia« of the an-

Anton Dohrn in his study at the Naples Station, November 1889

nelids. Over this point the most biting controversy with Gegenbaur began.

A wealth of developmental facts was brought to light that are of lasting value and that also force the comparative anatomical views to changes in their interpretations and that initiate new investigations. But, in accordance with his goal, Dohrn felt forced over and over again »to enter the field of hypothesis«. As he says: »I step into this field fully conscious of its hazards.« Later on, he separated more sharply observations and their hypothetical interpretations, and one finds more and more frequently restrictive remarks. »On my part, I still adhere to the annelid hypothesis . . . I cannot presume to anticipate all the wrong tracks on which the hypothesis might go. I consider it as inevitable that it will do so, but this I do not consider as a particular misfortune. In this sense the most vehement antagonism of viewpoints is an advantage for the final result, for it leads to renewed attempts to extend the field of facts and to deepen criticism« (1890).

Over and over again, something becomes questionable »and cannot be answered successfully until our basic conceptions of the ancestors of the vertebrates have assumed a more distinct form« (1901). Thus the »studies« become more and more a tenacious compilation of embryonic materials: »phylogeny« moves further into the distance.

In a general consideration of the fundamental procedure of phylogenetic studies, Dohrn makes use of a characteristic comparison: »It is like a colorful mosaic whose individual parts are spread on the table in great disorder. Only an experienced artist can put them together according to the originally planned picture by the use of preconceived ideas about the whole picture and by observation of the characteristics of the individual little stones. Very many of them have been completely lost, however. But if the artist succeeds in reconstructing the complete picture, if he paints it on a canvas and then arranges the existing stones properly, perhaps he may also become aware of some errors in his arrangement of the canvas and improve it according to the indications of the stones. And if he is sufficiently endowed with imagination, he may succeed in recreating the lost stones and reconstituting the whole picture.« Then follows the fundamental remark: »The richer and the more diverse the embryonic and larval stages in a phylum are, the more distinctly can its phylogeny be determined. The poorer and more monotonous, the more difficult it will be to ascend to a common ancestor.« Dohrn's embryological and phylogenetic considerations have a certain convincing force. It must, he claims, have come about in this way: in one case, a degeneration in the uptake of food must have occurred; in another case an advance occurred, under the pressure for more effective defense. Here, Dohrn makes use of a heuristic principle »whose general validity has hitherto not been sufficiently appreciated«. He formulates it as follows: »The transformation of an organ comes about by a sequence of functions within this organ. Each function is the sum of several components, one of which is the chief, or primary, function, whereas the others represent additional or secondary functions. The decline of the primary and the rise of a secondary function changes the overall function. The secondary becomes gradually the chief function, the total function becomes a different one, and the result of the whole process is the transformation of the organ.« With this »Principle of change in function«, Dohrn gave to comparative morphology, as it relates to genetics, a guideline for critical and combinatory observation. This thesis transformed some old opponents into friends but it also found strong opposition. It remained methodologically important to him.

Carl August confirmed the receipt of his son's publication with a pun taken from the annelid thesis and baptized the author as his »Anneliderjan«.* As with other things of this sort, his son put this aside. There was tension again between Hökendorf and Naples.

The correspondence with von Baer was thorough and amicable. He declared the annelid thesis as plausible if there had to be phylogenetic speculations. The principle of degeneration also had his »approval« and he »certainly had no objection« to the change in function. But he objected to Dohrn's speculation about the origin of vertebrate limbs. And he indicated clearly his basic disagreement »that I don't feel any need to derive the vertebrates from another group«. He would not deny mutability, he had articulated it before Darwin, »but I cannot accept Darwin's explanation of it, i.e., the selection theory; I am against it. It takes as a point of departure the idea or the sentiment that reason does not prevail in Nature. I,

* Liderjan is a disorderly person. Dohrn's father made a pun, deriving it from the annelids.

however, cannot break away from the old view that reason or teleology is in it, be it immanent or transcendental.« The conclusion of the great letter of June 11, 1875, is: »Whatever may come, we won't fight over Darwinism, etc. At any rate, by the foundation of an Institute in which the embryonic development can be studied carefully and the embryos can be preserved well, you have initiated a progress, that will keep a tight rein on Darwin's speculations, if necessary. If you don't find the reins in Naples, then they might not be necessary after all.« The 83-year-old man hopes to come to Naples before he »has to make the journey to the land of shadows«.

Dohrn became deeply agitated by Gegenbaur who, in 1875, published the first volume of his *Jahrbuch für Morphologie*. In the introduction he had described, as the essence of comparative anatomy, the elaboration of the comparative method as a substitute for experiment. But he emphasized that it had to be handled with the greatest rigor, »otherwise the procedure becomes uncritical and unscientific . . . A striking example of such an unscientific comparison is the well-known comparison of the so-called ventral nerve cord of invertebrates with the dorsal spinal cord of the vertebrates.« To the lack of discrimination is added »the most flagrant ignorance of facts . . . Such an unscientific comparison wanders as in a labyrinth in which one erroneous way is followed by others.« Dohrn's name was not mentioned, but it was precisely this that hurt his feelings, as is shown in a letter to Huxley, who was involved in this controversy. Every expert was aware that he was the target. And now began a fight which extended over two decades. Dohrn tried with attentive pleasure to seek out mistakes in the observations and deductions of Gegenbaur. He attacked him with a barbed pen and cold irony, his writing style is far superior to that of the somewhat brittle and boring man. In the middle of the 1880's, the fight assumed a somewhat grotesque form when Gegenbaur confronted their common publisher, Engelmann, in Leipzig, with the alternative: either Dohrn or me. Dohrn withdrew without pain. His adversary had played a new card into his hands: namely, he had turned the scientific fight into a personal one. This polemical controversy did not come to any conclusion as far as we can see, but it served to make clear the limitations and problems of the method, and this is never without merit. Dohrn might have felt that he was victorious, when he pointed out to Gegenbaur that comparative anatomy without embryology remains bogged down in description. Altogether, he treated Gegenbaur and Haeckel, who remained close friends, as a unit. But it is characteristic of him that when he was asked for his opinion about a nomination in Berlin at the beginning of the 1880's, he advised sharply against Gegenbaur and suggested Haeckel for consideration, notwithstanding the obvious contradictions.

The game of life brought about that the two former friends, who avoided each other in Germany, met again on English soil in 1898 when Cambridge University bestowed an honorary degree on both! This, however, did not lead to a rapprochement between them. Dohrn's journal entry of August 26 states dryly: »Then all of them went to the zoological session, where Haeckel preached, but Heape and I stayed home . . .«

5 ROBERT VON KEUDELL

When Rudolf von Delbrück, administrator of the Chancellor's office, said with a little sigh to the representative Heinrich Dohrn: »Keudell does everything your brother wants«, it was meant with some self-irony. He had himself become immediately interested (»He liked the daring of the whole matter,« Dohrn wrote to Stahr); his visit to Naples had convinced him that something very responsible was being done. But shortly afterwards, Delbrück resigned. That was a sign of the great change in Bismarck's economic policy. Dohrn's mother, Adelheid, whose letters intermixed in an odd way family stories, money matters, advice for household and education, with intelligent political reflections, had an idea about what this change of personnel could mean, both in general and in particular. She wondered how this loss of a supporter might agree with the adventure of her youngest son. He himself probably felt the same. So much depended on personal atmosphere. Bismarck had replaced the outstanding Delbrück with a colorless bureaucrat, Hofmann, from whom he had no fear of a personal opinion. When Dohrn had his first discussion with him, he noticed clearly enough how much the quality of his partner had changed: »I find the man obnoxious, his look is false, he is without principles and goals – a bad substitute for our old friend Delbrück.« This he wrote in September 1876 to Eisig. The office remained important, of course, particularly for the accounting to the Federal Council. To Dohrn, however, this official relationship began to change decisively and had many consequences. This development started when Keudell was offered the ambassadorship in Rome.

The new representative of the Reich was 48 years old. It was known that in the 1860's he had belonged to the closest circle of collaborators of the Prussian Prime Minister.[27] His early intellectual and professional development had not inclined him in this direction from the start: he had, as a young man, associated more with the literary and musical circles of Berlin than with political and official ones – and then he had entered the inner administration. The early relationship to Bismarck had more of a human than a political character. Johanna Bismarck took piano lessons from him, and Bismarck liked to listen to him. Basically, Keudell came from a more liberal intellectual world, but the bold politics of the Prussian representative at the Frankfurt Federal Assembly attracted him; and when Bismarck took over the Ministry, he called Keudell to his side as collaborator. The man who knew the world and had traveled much belonged, in those decisive years, to the most intimate environment of the leading statesman, although sometimes, for example, in the Holstein question, he was quietly in opposition – for a while he was expected to be the coming Secretary of State of the foreign service. After a year in Constantinople, Rome became his second diplomatic foreign position, of which he took care for 14 years. While the relationship to the Berlin center became less close, he acquired strong ties in the Italian capital, at the court as well as in the population. His stately, impressive appearance in the light-colored uniform of the Hussars* made him popular among the Romans.

In June, 1873, Dohrn first visited the newcomer in the Palazzo Caffarelli and was able to tell him about his state of affairs. The ambassador promised to come to the opening of the Station. That in itself did not mean much, but it was the beginning of an ever-renewed exchange of ideas, during which Keudell liked to discuss the plan to take over the Station as a Federal Institute that could be done either by separating the work tables from the individual German States and placing them on the budget of the Reich, which would raise them, or through a larger, permanent subsidy. At first, this suggestion met with a negative response from Delbrück. The problems of science remained a matter for the individual states, the budget of the Reich had no real place for such things. One hesitated to take the first step. And every consideration met with the limited contract between Dohrn and the City of Naples. It was valid for 30 years – after that time the right of reversion became valid. Therefore, the state of suspension of the legal situation persisted. There had to be a change before Keudell's intentions could have a stronger position.

A lucky coincidence eased the way. In April, 1875, the German Crown Prince had visited the Italian King and had used his stay in Naples to look at the Aquarium and the Station. Two years earlier, his good services had had a favorable effect at the Berlin Ministries. His walk through the Station had been completely improvised, but what he saw and heard impressed the visitor so strongly that he put himself at Dohrn's disposal once again and asked him to visit Berlin. More importantly, the next day King Vittorio Emanuele visited the Aquarium incognito, which obviously reflected the great praise the Aquarium had received from his guest. Dohrn, to his great amazement, received the Comthur Cross of the order of

* A Prussian regiment.

the Italian Crown. Carl August Dohrn had had the somewhat coquettish ambition of spending his life without an order, and Anton's early political radicalism had not prepared him for a career in this field. But now he might, not without satisfaction, hang this distinction around his neck. For he had learned a lot in the meantime – that he started his entry among decorated men at a high rank was understood to be an event highly desirable for the status of the Station. Numerous bestowals were to follow (letter to du Bois-Reymond: »The King bestowed on me, on his own initiative, by-passing the Knights' and Officers' crosses – straight out the Comthur Cross of the Crown of Italy. Keudell declared this to be unheard of in the Annals of this order. I am too profane to know this.« May 20, 1875). Friends in the Berlin office were pleased with the event, and his enemies found it surprising. For the young Comthur it was a highly important legitimation in all future negotiations in Italy. Most important was the examination of the contract with the city of Naples, which oddly enough was formally signed only now after the strange paragraphs of the original plan had been dropped. This examination had to be introduced carefully; but Dohrn found out how much the mood of the Municipio had changed. Its Secretary General, Cammarota, had from the beginning had a lively feeling for the personality of the bold and tough German, as well as for his enterprise, and he became a loyal mediator between the diverse interests. The elements of the contract remained, but the period of possession was increased from 30 to 90 years; as the beginning year, 1875 was chosen, and in early June, 1877, the Giunta [City Council] accepted this new arrangement. Anton Dohrn wrote in a happy mood to his father-in-law, who had put the dowry of his daughter into the Institute: »It seems that, if the administration is well done, even the grand-children of Boguslav can receive the dividends and the great-grandson can receive a birthday present paid by last year's salary. Therefore, the present owner of the Zoological Station cannot be accused of being a prodigal son on whom, one day, his great grandchildren will pass judgment.« Boguslaw was Anton's two-year-old son. The old

Baranowski had once said: »The Station is doing better and better, and his children worse and worse.« This would no longer be true.

For Keudell, the improvement in the contract situation did away with the main difficulty standing between him and his main goal, namely, the creation of a permanent connection between the Station and the German Foreign Office. With an energy that certainly pushed the daring and venturesome Dohrn forward rather than stopping him, Keudell made the good fortune of the Station his most personal care. This energy did not at all fit the cautious and almost silent, phlegmatic personality which Dohrn had at first noticed and which had caused, in German-Roman circles, Keudell's reputation of being, after Moltke, the most silent man. Science, in the narrower sense, did not interest him very much – it was even foreign to him – and his diplomatic reticence during their first meetings had made Dohrn somewhat insecure. But they came together on a human level because of a shared passion for music. Keudell was an eminent pianist and Dohrn an insatiable but discriminating listener; he knew and judged music well, although after the Jena period he had given up his own music making. The same preferences and dislikes united them. And when, in 1875, Keudell visited the Station, and had the inner structure thoroughly explained to him, he himself offered help.

This he gave officially in every way, but the official relationship turned into a human and personal friendship. Dohrn once wrote that he shared his vacation with no-one, except Grant, with such pleasure. It was a joy to win the ambassador for the beauties of Ischia. There they were together for weeks. Later, Dohrn once asked him how it happened that, in the early Roman years, he supported the Station so much. Keudell answered that Dohrn reminded him of his youth, when he, around 1860, had united his personal fate with Bismarck's enterprise – who was considered »mad« – although all his friends dissuaded him: »When I saw that you had devoted your entire existence, your finances, and your freedom to a great, impersonal goal, my inner sympathy was awakened.« Keudell's decisive contribution was the great memorandum to the Foreign Office of August, 1878, in which he supported a permanent subsidy from the budget of the Foreign Office. Dohrn had agreed with him that in case of a German subsidy not only the past financing should be examined, but the future should be put under permanent control and judgment of the Roman Embassy.[28] That meant desirable protection for himself against distrustful gossip. In a letter of July, 1878, Dohrn thanked the Ambassador, who had returned to Rome, »for everything that has happened and will hopefully happen in the future. According to Goethe, all human beings, according to their nature, pay either with what they have or what they are. I am afraid that both methods of payment would find me insolvent and I must depend on your finding joy, if the work succeeds, which I would like to give you if I knew how.« Keudell, he wrote, had made people grateful, »even if the attack does not succeed«. His feeling was one of hope, because he knew that he was on human ground.

6 THE SUBSIDY FROM THE GERMAN REICH

Dohrn and Keudell were of one mind concerning the goal of finding lasting support for the Station from the Reich. But they were both mistaken concerning the time during which it could be achieved and concerning the influence of the powers which were to be used. After Delbrück had rejected the idea of fully taking over the Station, the plan was simply that the Reich should rent the work tables, enlarging their amount to ten. That would have procured great security. Prussia, with three tables, Bavaria and Baden with one each, could be counted on as firm property. Now, a good basis had also been found in Württemberg. But the Saxonian Ministry was still uncertain – from there, a negative judgment of the Station had been sent out to the world, which had its effect as far as Berlin. The interference of Leuckart from Leipzig balanced that. Mecklenburg, which had shared a table with Hamburg, had won back its promise of half a table. The Grand Duke told Dohrn that one could talk about it again, if one of his subjects were interested. No budget could be made on this basis. Hessen-Darmstadt was taken as a substitute. But it remained a difficult matter to count on the atmospheres in the bureaucracy, and the uncertainties of parliamentary good will.

This conquest of Berlin cost three campaigns, with changing strategic plans and tactical information. An unexpected difficulty evolved when, at the annual meeting of naturalists, a committee was created by Kossmann and others – Haeckel also belonged to it – which demanded the foundation of Stations in Kiel and Helgoland.* Of course, Dohrn could not voice any basic objection to this. He also felt that it was a bad tactical disturbance when Frank Balfour, out of a lack of skill and too much eagerness, wrote in the English *Nature* a negative criticism of the plan. It was undesirable for the story to reach the imagination of the Ministries. His own idea had been to let Prussia take care of it – »my color is black, white, and red«. In any event, the Station on Helgoland was started only a decade and a half later.

Dohrn's main card was the kindly disposition of the Crown Prince. That the latter himself took the lead for the subscription of *Fauna and Flora* was welcome, but he was also supposed to become part of the financial decision. On October 5, 1876, Dohrn was invited to tea. About this reception and the Crown Princess, Dohrn wrote some amusing letters; with the Crown Prince he talked about Italian politics, and complete success seemed to be achieved in the matter of the Station. Friedrich Wilhelm promised his visitor a letter to Secretary Hofmann with the request that he treat Dohrn's application as his own. No more could be expected. But the bureaucracy felt that the withdrawal of Delbrück did not particularly strengthen the tendencies of the Crown Prince – Dohrn, too, had to experience this, since one year later he was again visiting Friedrich Wilhelm, who complained that there was no progress; he had after all lived through the beginning of the matter and had been

* Kiel is on the Baltic Sea, Helgoland an island in the North Sea.

able to help. »In a hesitating tone,« acknowledging his »lack of influence« (thus Dohrn formulated it in a letter to Marie Dohrn of October 20, 1877), he said: »I will try once again to start the fire.« But the fire did not burn. And when Dohrn made his final attempt in February, 1879, he wrote to his wife, who had just been visited by her father: »The Crown Prince has just been disarmed, he cannot do a thing. Tell this to your father to prove how republican we have become. Keudell is also powerless.«

There was still another line to the Court: Prince Karl of Prussia had been in Naples, and Dohrn had to devote much time as »bear-guide« and companion and was invited to Glienicke several times. As far as political action was concerned, the old soldier could not do anything, but the Prince had, with success, strengthened himself and obtained 10 000 Mk from the »disposition budget« from his Imperial brother – Wilhelm I had already been made aware of the Station by Moltke.

The difficulties of the Ministry were mainly a question of the competence of rights of each State; Dohrn was told that the Federal Council and the Reichstag would not cooperate, and the investigation of the realm showed the opposition of Bavaria. A trip to Possenhofen to Carl Theodor, the medical man, was supposed to break the resistance. The learned Bavarian Duke was not only polite, but interested. By accident his sister, the Empress Elisabeth of Austria, was present; since Dohrn hoped to gain something for his position in Vienna, he told her his worries and goals. All the Duke could promise was that he would see what he could do. He could not do anything. Secretary Lutz could speak only for his own country – but the Reich? No. Did the way *have* to go through the Federal Council? Who could hand in a petition there? Dohrn thought of Friedrich of Baden; his brother-in-law Wendt sat as director of the Baden schools in Karlsruhe and through him there were relationships to the Grand Duke. They were willing to help, but the initiative had to come from elsewhere.

As early as November, 1877, Dohrn found out that Bismarck
 would have no objection to the Reich's taking over the worries of

the Station. A question coming from the Emperor's staff had brought the matter before him. Dohrn greatly overestimated this report. On November 14, he wrote to his wife: »As you know, I am always lucky. Thus, I found the sympathy of the Chancellor, who is generally not too well disposed toward the sciences. With this, the last height of what I wanted to achieve has been achieved. Beyond that there is nobody.«

The comment that Bismarck had »taken care of the matter« (as he says in this letter) was in fact premature. That happened at a much later date and at that time in a not very agreeable way. Secretary Hofmann had expressly rejected, in October, 1877, »the idea of working toward a permanent contribution to the expenses of the Institute from the budget of the Reich«. He asked for a presentation of the income and expenses for the year 1876. Then one could talk again of a contribution out of the disposition budget. Keudell, above all, wanted to take the whole matter out of this situation and he decided, in the course of the summer of 1878, to get an exact picture of the inner situation of the Station. »I myself asked for this investigation,« Dohrn wrote to Siemens, »because I hope to get the best results from it«.

The new Consul General in Naples, Beer, whose bank was in contact with the Station and who, therefore, had a sufficient idea of the financial foundation of the enterprise, gave his advice to the Ambassador. »The result was very satisfactory,« Dohrn wrote to Siemens a few weeks later, on July 26, 1878, »the Ambassador is of the opinion that, if a corporation or the Reich had built and furnished the thing, it would have cost twice as much. Also, he found wages and running expenses as low as they could possibly be, without interfering with the development of the whole. Out of these thorough examinations grew the great memorandum of August 1, 1878, in which Keudell suggested that a subsidy should be taken over by the budget of the Foreign Office. It became a most important document for the development of the Station, because it formed the basis for the negotiations of the Offices, the Federal Council, the budget committee of the Reichstag.

Dohrn himself had started, in case this attempt should be successful, to create some kind of administrative council, consisting of scientists who had worked in Naples – they should make the effort to supervise the accounting and expenses.

This suggestion recurred also in Keudell's memorandum. But he opposed the creation of a formal control (as in the Archeological Institute), because this could hinder the development, which depended on Dohrn's personality. Perhaps it would be enough to demand of the directorship a yearly economic report – and to authorize the Embassy and the consul general to report regularly and to control the books.

In order to pay off the deficit and gain a certain freedom for the necessary development, Keudell suggested 40 000 Mk. After the professional and personal expenses had been discussed in detail, Keudell had arrived at that sum by suggesting that 2% of the basic expenses be put aside for amortization – separate from the deficit arrangement – and that Dohrn be given a salary of 3600 Mk for the directorship. For his own living expenses, Dohrn had up to now taken out the 5% interest of the money he had put into the Station. Keudell's explanations are thorough, and simultaneously plead very warmly for the cause and the man.

The Secretary of State, von Bülow, did not like this very much. Dohrn, who was in Berlin in September, reported home a comment of his: »The Foreign Office does not need zoologists.« Keudell, too, did not make any progress, brave though he was. The Crown Prince was approached again, but von Bülow gave in only hesitantly: he told Dohrn that he would not go past 15 000 Mk.

The year of 1878 was a year of great inner political restlessness and uncertainty. Dohrn was by that time an outsider, he looked at events more calmly than in his years of passionate drive, but they still reached him and his work; he remained a sharp observer. It characterizes his sharp sense of combination when he writes to his sister – after the first of the attempted assassinations on Wilhelm I which followed each other in that year: »Will Hödel cost us Falk?« Does the time of Liberalism end or will it truly start, when

Friedrich Wilhelm follows his old father on the throne? The impressions in Berlin reflect this uncertainty. There is some talk that Bennigsen will become Vice Chancellor, perhaps even the successor of Bismarck – Dohrn's letters make a note of this, although he did not yet know how to evaluate it. When the Secretary of Education, Falk, who favored him, was dismissed shortly after the attempt at assassination by Nobiling, this was a clear signal for change.

Dohrn's fear for his own cause, however, soon proved to be without foundation because the succeeding Secretary of Education, von Gossler, supported him strongly, as the next years were to show. This he did not know at the time. His personal relationships belonged essentially to the national liberals and free conservatives. Obviously, the task was to win a broader basis in their sphere. There was one difficulty among the conservatives, which Dohrn considered odd; the representative von Behr-Schmoldow had created for himself a kind of monopoly for everything that had to do with fish, but he found that the Reich should not have to worry about the fish in the Mediterranean. The question of biological research did not move him. It was Keudell's task (and the representative Count Moltke supported it) to create a friendly disposition in this group. The problem was more critical among the Catholic Party. Dohrn had been warned earlier, shortly before his official programmatical speech of 1875, not to emphasize his Darwinian point of view; that would cause him troubles. Of course, he rejected this advice. He strongly underlined his idea of how much biology had to thank Darwin for his decisive stimulation; but he explained at the same time that the task of the Station was not to anticipate potential knowledge in terms of popularization. This speech, not directed to anyone in particular, proved to be useful when he had conversations in 1878–79 with the leaders of the Catholic Party, Peter and August Reichensperger. It found the unreserved approval of both brothers. Dohrn always gratefully recognized that after the clarification of scientific objectives, the Catholic Party unanimously supported his enterprise – although

Windthorst »did not like to do this«; he would have preferred an enterprise in Germany.

The most important thing remained Rudolf von Bennigsen's willingness to lead the parliamentary campaign himself. He had been in Naples and he, too, was strongly affected by the daring enterprise of producing, on such a dangerous ground, a work which, precisely in its international aspect, represented quiet but unmistakable proof of serious German scholarship. Dohrn's brother Heinrich became the middle-man to the party leader; that he held back was a matter of course and also good luck – Anton was not certain about the tactical skill of his older brother. The sometimes sharp sarcasm of the witty as well as self-centered man might cause problems.

A very skillful way was found for avoiding the possibility that the matter might fall into the hands of the parties – perhaps it was Bennigsen's idea. They chose the way of a petition outside of parliamentary circles. These »petitions« generally led a Cinderella existence in the parliamentary workings; their committees were often nothing but horror cabinets where the members were self-important and quarrelsome. But now a petition was put before the German Reichstag on March 6, 1879, asking for a subsidy for the Station in need. The piece was an eloquent consideration of the research method started by Johannes Müller; it said »without exaggeration« that the Zoological Station had raised the study of animals to a new level of development – it was signed by Helmholtz, du Bois-Reymond, and Virchow. The joint appearance of the three men of international scientific fame had to make an impression. When, on March 8, the budget of the Foreign Office was to be discussed, including Title VI, 13 – »Support of German Schools and other patriotic enterprises in foreign countries« – representative Römer asked that the decision should be postponed and the recently received petition should be transferred directly to the budget commission for consideration. That was done. The Foreign Secretary, von Bülow, was not very happy. He emphasized that he had carried out earlier interventions »with the greatest pleasure and complete willingness«, but now he could not hold back a certain caution: »without wishing to hinder the course of events and of the consensus . . . The matter is that the Foreign Office does not really, if I may say so, represent a permanent marriage with this Institute, but only a passing relationship.« The Reichstag received this joke cheerfully. Dohrn acknowledged it with a polite letter, saying that the relationship might well lead to a marriage; on March 19 the budget commission unanimously supported the request of their secretary, von Benda, to transfer 30 000 Mk into the »Extra-Ordinarium«. The handwritten protocol amusingly reflects the fight over the amount. First the words »at least« were written in front of the amount; then these were struck out and replaced by »if possible«; and the final version was »approximately«. Bülow had asked for caution and talked of 15 000 Mk, whereas Keudell's memorandum had counted on 40 000. He was somewhat afraid that the »responsibility and surveillance« would fall into the hands of the diplomatic representation – the Reich would hardly be able to »fulfill all hopes which are tied to this beautiful work in its first few years and its first eagerness«.

In the meeting of March 20 certain doubts were expressed, particularly by the free conservative Lucius who »suspected« in this unusual petition a break with the usual practice – this was a matter for Italy. The reservation must be expressed that this was a grant to be recalled at any time »if the scientific results should not correspond to our expectations« or if the administration should be inappropriate. Bennigsen, Windthorst, Hänel, Maltzahn-Gültz, and Lasker entered the debate: the national liberal reporter von Benda talked against Lucius: the idea that it would be better to sell the Station to the Italian government should not have been expressed in that room. In general, however, the debate was very friendly toward the Station, and Hänel above all stressed the fact that the »unusual case« demanded freedom from previous examples; the doubts that came up referred to budgetary consequences.

Keudell and Dohrn were only conditionally satisfied with the results. The hopes of getting put into the ordinary budget immediately had been too optimistic. For that, certain presuppositions, such as the judicial situation of scientific officials, had not been sufficiently clarified. Bennigsen made, on March 5, 1880, the attempt to get the Station into the regular budget, but Freiherr von Stumm and the Secretary of State opposed this, saying that this was »not yet secured«. It was a disappointment that in the accepted amount 10000 Mk were missing. The financial worries of the following years in running the establishment often made Dohrn sigh that Virchow, who as a scientist had always been on his side with advice, recommendations, and personal participation, had a tendency toward a pedantic budget parsimony when he was representative of the Eugen Richter group. In the budget commission he had suggested the sum of 30000 Mk. How much writing and running, how much reproach and distrust Dohrn would have been spared, if the decision had turned out differently!

In 1889 the goal was finally reached. In the meantime, Dohrn's and the Ambassador's reports had been received regularly, and in 1881 they appeared even with all individual accounts among the printed matter of the Reichstag. Now the budget commission had not only a clearer idea of the achievement and the bookkeeping, but the Station had actually entered the consciousness of the nation on a secure footing. During the budget council on February 6, 1889, Bennigsen discussed the situation: »The number of tables has doubled, over 500 scientists have, up to now, worked in Naples, half of them from Germany.« Since work and administration costs had grown, he asked whether the subsidy should not be enlarged, »so that the Institute can be secured beyond the life of such a man«. This request had been arranged in advance, and the Secretary of State, Herbert Bismarck, thanked Bennigsen: »I have repeatedly visited the Institute . . . It is, in my opinion, one of the best institutions of practical science which exist anywhere . . .« Only because they had not been certain about »the mood in the country and in the Reichstag«, had they done nothing further in the matter. That should be done now. Obviously, the speaker was not quite enlightened about *where* the difficulties had for some time been located. That they were now being set aside alleviated Dohrn's worries for a span of some years. As always, however, the relief was only for a span of time – his inventive mind was already envisioning the next goal. He turned forward towards it with an intelligent estimate of his powers, and with the instinct that somehow and at some time he would be able to master the difficulties.

7 MARRIAGE AND FAMILY

Dohrn's life was based on risk. He had undertaken a work without predecessor or comparison and had fought his way through obstructions and disappointments. There was a restlessness in him, an obsession with his task which, seen by him early in its great contours, grew constantly and captivated him even more. At times, he was its driven slave, when he should have been its master and had to become its free master. In between there were the pains of depression, the tiredness of not being able to do anything, and the fear of not being able to wrench from fate the proof of the intelligent and bold hypotheses, which were born in his deductively working imagination, while he was sitting over his microscope or microtome. He was never worried about being defeated by his efforts – he was fundamentally of a dominating nature. But after having felt the first disappointments to be a true defeat, he knew that the fight would demand the entire man. Therefore he had included the unmarried state into his program. He himself felt his marriage to be daring, after an almost paternal friendship unfolded into love, although it was a great happiness.

Even greater was the risk for the 18-year-old Marie von Baranowska, but she was not aware of it. Since her childhood days she had admired this man, who had entered her parents' home as a foreigner and who, with his ideas and plans, with his free, grand manner, had moved everybody, not least her mother, who now fought so much against a connection with Dohrn. But she had lost the fight. The arguments against the marriage paled, since the previous close friendship with the family had been a proof of their mutual esteem. The young woman may have believed that, seen from a broader perspective, her life would not be very different from her previous life. She would continue to live with mother and sister in Palazzo Torlonia, the same old friends would come and go, there would be the old conversations, excitements, triumphs . . . and that her husband would still have to take his trips several times was unavoidable. That would soon be over, and then it would be marvelous to participate in the creative development of his work and being.

Into the first years of the young marriage – they were both mutually thankful for their happiness – there quickly fell, in fast sequence, the shadow of pain. Katharina von Baranowska, whose delicate health had dictated their stay in the South, died early in January 1875. The sister, Helene, had not been well for some years. Charles Grant had a shy love for her; it was hidden in delicate verse in which resignation ruled, since feelings were barely expressed – the girl became engaged to a Polish refugee. Anton had to accept the resignation of his friend: »He is very calm and resigned. I do not know how he feels inside himself. A lot of heart's blood might be spilled. He does not say a word about it . . . Grant feels crucified in favor of the first man who finds Helen pretty and interesting . . .« An almost passionate argumentation turned against the planned marriage. Dohrn felt partly responsible; he did not want romantic stories with a very uncertain background and he warned Marie, who wrote to him about her sister's happiness: »You *must* not mix up the symptoms of the happy, fulfilled ›being-in-love‹ with the happiness of ›married‹ love.« This event, which strongly and frequently moved the writer of this letter, took place in the summer of 1876; in January, 1877, the girl died at the age of nineteen. Her brother Casimir's death followed in May of 1879 – he had always been a source of worry and loving care, and was staying in Belgium for his studies at the time. Since death was near, the tubercular man went to Naples. He could not be helped. His father came while he was still alive. Dohrn was away on a trip, and

Marie Dohrn, née Baranowska. Naples, around 1885

Lobianco took the corpse of the young man to his Polish homeland – this act of love connected him even more closely to the family. This sad trip was the first trip into the world, which the young Italian, himself so full of life, had to undertake.

Dohrn was traveling much too much. From his wedding, celebrated in Warsaw in 1874, he traveled immediately, alone, to St. Petersburg and Moscow, which had to be visited personally because of the work tables. Then he took his young wife to Hökendorf to meet his parents and siblings. That went well; his mother was happy to know that her youngest son was happy. Her household worries concerning the new daughter, who had barely outgrown the age of childhood, were somewhat strenuous, but not too heavy. Carl August showed off his art of being a gentleman. He enjoyed playing homage to the charming woman, with her large, somewhat astonished eyes, in all the languages that he commanded and she understood. Marie herself was not yet completely fluent in German. This flaw, which later disappeared, made her brother-in-law Heinrich her victim. Marie translated »beau-frère« with »Schön-Bruder«* and thus Heinrich always kept the nickname; although his narrow face with the searching eyes and his somewhat untidy beard were characteristic, they could not be called beautiful.

In their early years together, Marie Dohrn participated directly in her husband's scientific and organizational works. Her original preference belonged more to literary and historical matters. But now she cooperated happily, taking care, for a while, of the *Journal* of his correspondence, and making copies; but she also entered the laboratory (»Now, finally, I can work with the microscope five to six hours a day, and have the clear-eyed woman as a charming assistant,« Dohrn wrote to Stahr on January 19, 1876).

They lived in the old apartment, in a somewhat improvised way, with the old furniture and kitchen utensils; but it turned out that the humidity which the rock behind Palazzo Torlonia produced

* »beautiful-brother«. Her German is a literal translation of the French beau-frère. The German word for brother-in-law is Schwager.

became a problem of hygiene. The tiresome search for an apartment finally led to Portici, which the doctors recommended. There they had an unexpected opportunity. The Palazzo Reale, an old Bourbon castle, was empty and costing the Province, which had taken it over from the Royal Private Budget, money for maintenance. The province was grateful to rent out some rooms rather cheaply. Dohrn agreed to the idea of living like a bourgeois in those halls. It was an odd way of living, full of contrasts: the furniture and the means for a decent living were absent. A beautiful park, with many trees, gave dignity and comfort to this romantic adventure.

But this gain was bought with the inconvenience of distance. The pretty living quarters were six kilometers from the Station, and the means of transportation were mediocre. The young family life was more or less torn apart. When work was urgent or visitors tied him down, Dohrn could not make this trip four times a day. Often enough he used his quarters in the Station. Fortunately, there was Grant, who also moved to Portici. The common grief about Helen tied him to Marie Dohrn, and he shared her literary interests. It is typical that Dohrn once wrote, somewhat worried, that Marie should not let herself be led too deeply into the English language, that Goethe was more suitable for her now. He realized the necessity of winning the mother of his children totally over to German intellectual life. Quite unconsciously, he had, out of the fullness of his own nature, evened the way toward this in the years of being her amicable mentor. At the same time, he was very thoughtful – with a natural respect for intellectual growth – not to tear his wife away violently from the elements of her inner heritage. He read Polish history eagerly in those years, and his letters to her show the lovable features of the noble Polish passions wherever he found them, but also the unfortunate elements of the political leaders of the Polish people. He showed her the charm of historical colorfulness, but also the weaknesses, the lack of a great, superior person. While he himself gained clarity and insight, he carried on a quiet education.

But he was tortured by the knowledge of owing much to his wife in those early years because of his unceasing travels. She did not reproach him, she knew too well what was needed for the enterprise, but she felt some disappointment. It was a difficult task for the 20-year-old woman, demanding much self-denial. Dohrn wrote almost daily to report the details of what he did, and to demonstrate to her the necessity of decisions. Simultaneously, he repeatedly assured her that the financial troubles would soon be over. Their house would stand on the Posilippo. It would see the common future, its name would be »Villa Baranowski.« After the age of 40, there would be an end to traveling, and then the steadiness of quiet and secure work would arrive, for which she would be indispensable to him. She would be his »saving genius in whom I will find love, understanding, forgiveness – everything that Grant, in his excited periods, tried to find in the Catholic Church.«

Marie Dohrn gave five children to her husband: Boguslaw (1875), Katharina (1876), Wolf (1878), Reinhard (1880), and Harald (1885). Naming them occupied him very much: the »program« which he repeatedly talked about with the young mother and his father-in-law was to alternate between Slavic and German names. Obviously, he moved away from this for his younger sons. While the father faced tough discussions in Berlin, a little girl was born at the small summer residence of the Baranowskis, Kazimierzofka, near Warsaw, in October, 1876; but in July 1877 she had already died in Portici. One cause may have been the insufficient medical care which the suburb had to offer. This loss hit the young mother very hard; and she felt it strongly in later years since she had to give away her sons, one after the other, because of the German school system. The question of a change in living quarters was all the more urgent after this death, but it dragged on for a long time. Apartments for rent were either very expensive or they did not correspond to the sanitary demands which were so urgent for the strangers after all their experiences. They searched for years to find a fitting place. The search was mixed with the plan to create a kind of off-shoot of the Station on the Posilippo.

Was there now a different atmosphere in Hökendorf, since it was no longer a question of the »painful Station«, but rather about a »real« establishment? The circumstances were not favorable at that moment. The oldest brother, Wilhelm, who was a better musician than businessman and landowner, went bankrupt. The reports that he had been taken in financially when he bought his estate make a somewhat doubtful impression. Carl August had to help with large sums and keep the family of five children until a new existence was secured. Wilhelm entered the commercial insurance industry. The beginning was not grand; but a certain winning personality and energy helped him and after some years he created, with skillful activity, a new distinctive position. But only after some years. In the beginning, he had before him a moody father who evaded business and was angry that anyone came to him with such a disturbance. The business situation remained critical in the second half of the 1870's. For a long time, the organizing procedure which followed the feverish mood of unification after the France-German war, dragged on. In 1877 a bank in Stettin went bankrupt. The Dohrns were losing 100000 Marks. »Bad luck,« Anton said in the note he sent to Eisig. His mother sent a summary of the situation. Of the estate which Heinrich Dohrn had created and which they had left to fate, there still existed several hundred thousand Thaler. But some nervousness remained after the misfortune. His mother, too, advised against larger investments at the end of the 1870's. She had heard that Moltke was going to ask for more soldiers – that meant more taxes and perhaps war.

For G. Baranowski, however, a great financial burden had been removed: he won a suit against the Radziwills, which meant 50000 Rubel for him. He used the sum to buy back from his younger brothers and sisters the old family estate, Wydranka, in the region of Mogilev. His son, Casimir, was supposed to learn how to manage it, but he died in May, 1879 and now Marie Dohrn was the only heir. In a quick decision, she offered to use part of her father's finances, which Baranowski could spare (in addition to her dowry which was already invested in the Station) for a mortgage for the running bank debts. Dohrn rejected this. He was anxious to avoid the mixture and confusion of the financial matters which threatened since he himself had no definite income. His goal was, rather, to remove Baranowski's money gradually from the Station. His father-in-law, who was living in Moscow and was just learning to think about his grandchildren, at times had an uncanny feeling concerning the whole matter toward which, years ago, he had felt a supporting sympathy. Not that he retired. During the Turko-Russian war in 1877 he spent time as the director of a hospital organization in Romania, where he made contact with Prince Karl who had just fought the »High Gate«* to win independence for his country. He immediately tried to interest the Prince in the Neapolitan Station. A few years later, this bore fruit. Dohrn himself had met the Prince of Hohenzollern during their common time in Berlin.

During the Spring of 1878 Anton Dohrn had visited his parents to find out about the situation and, as far as possible, to gain some clarity. That apparently failed completely. The relationship between father and son seemed to break down. A bitter letter to his sister in May, 1878, complained about the role which he always had to play in Hökendorf – he was still treated as the youngest son. He always had to apologize that he had dared to win the approval of the most significant men and the envy of the most insignificant by an action which grew by contradictions: »You will hardly know that I now count among those whose heritage has been reduced to the legal portion; probably as a consequence of my disrespectful expressions against the use of several unparliamentary words against my person.« It is also possible, Dohrn goes on, that he was angry only about his success, which puts him in the wrong when, full of distrust, he advises him against his plans. His mother saw it so, and she suffered from this disagreement. But in her letters, even

* Turkey, i.e., the former residence of the Grand Vizier at Istanbul, until 1918 seat of the Turkish government.

when they admonish her son to be more considerate, she took a position against her husband. »Tell Anton,« she wrote to Marie, »that property made his father harsh.« She also wrote, »he cannot be a tender father, he is a despot«. The will, she thought, was written in anger and was not valid if the children agreed simply to share the inheritance among each other. Anton was deeply hurt that he still appeared to his father as an »immature visionary,« as a »miserable dreamer of projects«, and also by his brother, who reported to him the varying opinions of his father. The threatening discord of the brothers oppressed the old mother most. »Zoologists are also human beings, just as entomologists are,« she wrote in December 1879 to her daughter-in-law, »but they are always boring for women, because the conversation rarely deals with anything other than the animals that are being murdered. Heinrich is a good man, but different from Anton, and Anton wants him to be different, instead of accepting him as he is. We all have flaws and those of Heinrich are that he never answers . . .« The women, she felt, should hold together and help make the peace which was so necessary.

But the tension continued. »Before I write to this man,« said Anton, »I would rather have my hand cut off.« This alarmed his mother, and she wrote in the summer of 1881 that he should not »argue man to man« against the 75-year-old man. And then she adds a rather grim, but nevertheless moving analysis: »What your father has against you is nothing but jealousy. Your name is mentioned, your diligence, your work is recognized – and that has no access to his heart as it has to mine. In large circles he was celebrated as a singer of songs, he is *President* and likes to be praised often, as you like to see yourself in print. It makes him happy, he brings me any little notice; what if I had not been – he thinks – and likes to hear it often. He is a weak man, but nobody is allowed to notice it; he craves recognition which must not stop. Be patient and indulgent, so that you may, once he exists no more, be satisfied with yourself and calm. This is my request.« This made an impression. He dared to pay a visit and repeated it (»In August father exploded, but in September he was sorry about it and he was gentle and friendly,« Dohrn wrote to Werner Siemens). His mother was happy that the experiment was successful through the self-control of her son and his prolonged stay. »I will never forget this.« She had reason to hope that she had brought about peace when she died two years later. Death came to the 80-year-old woman in September 1883. Anton wrote to Eisig at that time that he was glad for her sake, but that he himself felt the loss very much, »all the more since she was a mother in every sense of the word«. She had said about Anton, »I never heard an unkind word from him«.

What was preserved between father and son was officially called a »modus vivendi« (way of life). It remained fragile for a long while. Some months after the death of his wife, Carl August wrote these cool sentences to Marie Dohrn: »Anton must know that, after what has happened, we can talk only of a ›modus vivendi‹ between us. His interests are not mine. If he makes his plans he should leave me out. I can only say that the manner he uses to procure his means of execution are not after my heart . . . But these are, as I said, his concerns, not mine. I only do not want to be burdened with them.« At that time Dohrn had new plans. A large laboratory boat was supposed to be built, financed not by public means but by private enthusiasm, which he believed to be effective. In Stettin they first remained cool, but gradually a means of approach emerged. Hökendorf was treated with slow caution, Hökendorf »where, as is well known, liberalism rules only as long as the individual's demands are satisfied« (to Anna Wendt, June 1885). An ironical report to Eisig from August of that year says: »The cigars are good and the old man let me have them freely, strangely enough.« This is a sign of friendship and confidence – for cigars, a definite kind for definite hours and occasions, were almost the only enjoyment in Carl August's way of life – his cigars were cared for with great knowledge. It cost a lot of time »to keep the old man in good humor – but I am after all not here for my own pleasure«.

Then there is a turn of events. If Adelheid was correct when, some years earlier, she had written about Carl August's conflict with his oldest daughter, namely that »he always had to live in a state of war with one of his children«, Anton was now in the background and it was Heinrich who, in somewhat lamenting letters, felt himself to be the martyr. He could not get along with the niece who now took care of Carl August and who participated in his musical joys and his travels. The old man was indestructible; now he traveled each year for the winter to Palermo, where he had discovered an Italian host with entomological interests. When, on that occasion, he visited his son in Naples, whose Station had in the meantime doubled its size, without his being »bothered« with it, he was finally satisfied. Now he wrote again to his son, and although he could not avoid being ironic – a worried letter of Anton »is like a new edition of the Book of Job« – everything became milder. Perhaps he might even change the angry decisions of the year 1879. Heinrich wrote in June of 1888: »I believe that he feels shy, out of superstition, faced with the writing of the will – quite aside from his dislike of all business activities.« But finally he arranged the matter; the anger was forgotten and the will revised.

The conflict between father and son, which rose and fell for almost a quarter of a century, has a more than private significance because of the range of matters at stake and the strength of the two tempers opposing each other. It was also a matter of finances because, in a strange way, neither of them had a business sense. Carl August's relationship to finances consisted of an odd indifference; he was satisfied with having inherited an estate from his active and able father and he was not interested in augmenting it by his own work. Whatever he spent for collections – a considerable amount – was never meant for speculation, but from early on, as donations for the general good. The administration of his estate C. A. Dohrn left to his wife – which was the most intelligent thing he could have done; and later to his oldest son, which was not so wise in terms of the results produced. But Anton had a subconscious feeling that contested the right of the father, who had simply inherited his wealth, to hold back sums which would help his son. Thus it happened at times that, in his opposition, he became blunt and harsh. He once used the hard words that the family lacked »ethical weight«. His father appeared to him to be too little tied to society. His father's naïveté, which grew from the wealth of his talents, contrasted with his own reflective self-conscious nature which was always conscious of his actions and justifications.

But in all conflicts he preserved a firm image of the strange and great powers of this man. That appears in the words to his mother who at that time (1883), close to death, had written about the aging of her husband: »I can never imagine father as an old man, because he represents so much the strong man who is superior to the usual weaknesses of life. Each time I come home I experience the purest joy of seeing him again untouched in the middle of his interests and his intellectual world. But it is precisely such natures who dislike to face the possible end which the irrevocable presence of death creates in their environment, just as they are not curious to consider their own life, and wait for the end quietly. When father was so sick in Bagni di Lucca and really expected to die, he made the impression on me that he would like to hide in order to die alone – just as one tells about the large beasts of prey that they hide deep in the jungle, in order to die there.«

With the distance of the years, having become older himself and finding himself in view of the uncertain future professions of his own sons, Anton Dohrn learned to see his father's attitude toward him more freely and to replace his stubborn defiance by sincere gratitude. His manner did not become sentimental – that would not have suited »C. A.«, but he accepted the unusual character with secret pride and treasured it with new understanding.

8 ITALY'S CONTRIBUTION[29]

With the safeguarding of the Reich's subsidy, Dohrn had won a decisive battle. Being full of ideas, he was determined to use this position as a lever as soon as it became possible to follow up the programmatic growth of the enterprise. But he also saw a certain psychological flaw. Up to now he had gained the support of foreign scientists and foreign governments because he was considered a private entrepreneur. That might change, when the Station took on a more public aspect. Also, it had to be calculated that the position of the German government and parliament might become a rather impressive argument for foreign scientists to suggest the same approach to their own States. That was already true in France; England and Russia were developing plans for their own Stations. From the beginning, the leading biologist in Sweden had undermined Dohrn's propaganda despite the Court's support because he aspired to the same goal. Dohrn looked at these movements with divided feelings: they confirmed the fertility of his first endeavor, but they might weaken its central importance. Above all the connection with England, whose young scientists he so greatly esteemed, must not be destroyed.

How he judged the tactical problems becomes clear from a letter he sent to Thomas Huxley on March 25, 1880: »The historical-philosophical section of the Academy has achieved the great archeological Institutes in Rome and Athens; Curtius managed to achieve the excavations in Olympia; but the physical section has nothing of the kind. Then there arose the Zoological Station and its support among the scientists of the Academy. With this as a lever, I disposed of many difficulties. The more we achieve in Naples, the more reliably the ›patres conscripti‹ of the Academy will dispose of their influence and keep it. But there is also the danger of seeing the Zoological Station become an institute of the German Reich. That possibility I will find desirable only when I will see no other possibilities of developing it with full force. Right now it is very difficult to create and maintain an international enterprise. But it seems to me that a good beginning has been made and I feel justified in asking for the support of the important factors outside of the ›fatherland‹ as well.«

In the previous year, the British Association had, because of lack of means, rejected the petition to give money to the English scientists for traveling to Naples. A lucky, timely accident made it possible for Dohrn to balance the defeat of his friends. In 1879 Charles Darwin had received the Bressa Prize of 12000 lire from the Royal Academy in Turin. He wanted to donate the money to some impersonal purpose, and asked Dohrn in the spring of 1880 to acept 100 pounds sterling for the purchase of apparatus. Dohrn was moved by the kind disposition of the revered man and countered with the suggestion of using the sum for English travel stipends. Darwin of course agreed. Wilhelm Siemens in London, who understood and approved of Dohrn's tactic, increased this modest amount by a yearly donation of £ 50 – thus, in case the English interests should threaten to diminish, a psychological counterweight was launched.

A few years later, when the Italian participation had been activated, it became important to clear up Russian hesitations. Russian science was being wooed by the French. Therefore, Dohrn himself traveled to Moscow to inspect and support the cause. He wrote to Eisig that his action »has put an end to the competition of Villafranca (the French Station).« At the same time he addressed the Russian Secretary of Finance with the characteristic sentences (June 17, 1884): »Germany and Italy are only primi inter pares [the first among equals]. The Zoological Station does not belong to any nation, it belongs to science. And although a German founded it and its place is Italy, it still remains the useful property of all those who take part in preserving and developing it.« When these words were written, the decision in Naples had already been made to show the contribution of Italy to the Station in a new light. Just as the accommodation of the City of Naples to prolong the contract freed the way for a subsidy of the Reich, the latter had become an argument with which one could petition the Italian State to allow an extraordinary support. The personal position of Dohrn had become secure within a few years – his unusual enterprise was understood. Even if, in 1873, his requests for freedom from customs charges for machines and apparatus had been rejected, he could now, in the summer of 1876, win a fight concerning local taxes which freed him from the threat of having to pay 14000 lire.

The personal relationships became friendlier. Although in 1877 he lost Professor Panceri, who had loyally supported Dohrn's first efforts, his successor at Naples University, Trinchese, followed the line of his predecessor. Also, practical men came closer: in the summer of 1876 Dohrn was asked to become a member of the Administrative Council of the *Piscicoltura Italiana* [Italian Fish Culture]. »I did not accept this«, he wrote to his wife, »I do not want to have any dealings with commercial enterprises. If the gentlemen need my advice, my help, and my influence, they can have it and there will be no lack of counter-achievements. But that is enough. Although I would make some money as administrative councillor, I hope we can get that also in some other way.«

This reserve concerning internal Italian questions on personnel and other subjects was chosen with great deliberation. But wherever possible, Dohrn profited by the suggestions and wishes which confronted him. The fact that they wanted to gain his help for their fishing problems caused him to make the living conditions of fish into a particular research problem. The Department of Agriculture followed his suggestion and rented a work table for that purpose. This broadened into a larger participation of the South Italian provinces, all of whom were interested in fishing problems. Dohrn could for a while follow up the idea that courses for gifted young fishermen could be connected with this special department, which was equally fruitful for practice and scientific purposes.

He used the same clever tactics regarding the navy. The latter had always been helpful, and Dohrn's report to the Prussian Academy of July 23, 1877 tells about its obliging accommodation: the boat of the Station may anchor in the war port and get coal from the State's depots. All coastal stations were asked for support, etc. The Italian fleet even allowed the Station to use their diving apparatus for a long time. Dohrn himself was the first to undertake experiments with this equipment in 1878. His correspondence with Siemens dealt for a while with the special question of how this helpful instrument, of which he hoped much for the future, could be equipped with light and telephone. The Department of the Navy enjoyed this contact. When the Italian corvette »Garibaldi« was chosen to sail around the world, Dohrn was asked to name a young scientist educated at the Station to go along on this trip. A certain scientific ambition had become the international fashion among the larger sea travels. Dohrn refused this petition. This was a matter of principle. He had once emphasized to Huxley that he never gave »information« about the work and achievements of the young scientists. That would have put him in the Devil's kitchen. The Station was meant to be a free research institute and should never, even for the younger generation, have the character of a teaching institution with something like certificates. Admiral Bucchia realized, although reluctantly, how painful it could become for the German to make a decision among the young Italians concerning a much desired position. He himself had to count on these young, unbiased men with their purely scientific interests. But – this is typical of Dohrn – he turned the matter around: he suggested that the Admiral should give up the idea of a learned man who, on a warship, would always be in the way; he should rather send a young, intelligent officer to the Station. This man would then be trained in collecting and conserving sea fauna to such an extent that his work would probably surpass any scientist's. That made sense, but a change of secretaries postponed the execution. In 1881 a lieutenant, Chierchia, was ordered to be trained. The yield he brought home from a three-year voyage of the »Vettor Pisani«, together with the stimulation his work had meant for officers and men, were so impressive that the Department of the Navy rented a specific work table for itself.

Of course, Dohrn did not omit to inform other Navy Administrations about the Italian example. He asked his brother Heinrich more than once to try to get an answer from Admiral von Stosch. It was one of Dohrn's disappointments that Germany sent a Navy doctor only once, and did not take up the matter seriously; but only two decades later there were occasional scientific commandos. The example of Italy had an immediate effect in Russia and Spain – the

Departments of the Navy for both countries made contracts for work tables in the Station in the 1880's. Dohrn found it refreshing to have some officers participating among the international scientists.

Although the local mood in the early 1880's was altogether favorable, it might still be too daring to appear in public with the plan of an enlargement of the Station. The increase in work tables had made the question of space precarious, since their usage changed for climatic reasons. For a while, Dohrn toyed with the plan of creating at the Posilippo a kind of off-shoot of the Station. Good housing must be found for himself and his coworkers, as well as a chance for accommodation for guests; a small colony of the Station with a harbor and some equipment, and above all a physiological laboratory. Werner Siemens, who was at that time Dohrn's special confidant, became a little worried about these intentions. With all the good development of the matter, Dohrn could barely get out of his current financial difficulties. Dohrn calmed him: »You do not have to chain me down, as you threatened. You can deduce this from the fact that I, although with a heavy heart, have given up the Villa on the Posilippo. It is too expensive.«

But his resignation only postponed the necessity which threatened. The precedence of the Station must be preserved by making new methods possible. The danger was great that the French would defeat him in this. These were after all not new things he had in mind, as he said several times. Already in his programmatic essay he had talked about the inevitable step into physiology, and on July 7, 1882 he presented the situation to Werner Siemens: »The physiologists have the lively consciousness of having reached a point where nothing is left except dead leaves and where they barely represent an independent discipline, but are becoming the maid of medicine. They are hungry for new tasks. A courageous person, Dr. Krukenberg of Heidelberg, has followed up the idea and investigated several species of marine fauna chemically and experimentally. Suddenly everyone realizes that a boundless mass of new problems is raining down on us. If I now create a physiological laboratory, a complete change will gradually take place.«

At this time, with his eyes on Germany, Dohrn had already prepared a new project which will be discussed later. But he also prepared for Italy his plan for the physiological laboratory. He went about it carefully. The first question to be solved was the matter of location. He thought of suggesting one of the houses still standing in the Villa Reale, for the proximity of the new work places to the old administrative center was technically desirable. But even his friends advised him to drop this idea. The memory of the storm of public opinion in 1873 should not be awakened in the Municipio. Therefore, Dohrn realized, it became a question of promoting the idea through public opinion.

When he heard that London planned a fishery exhibition for 1883 similar to the one in Berlin in 1880, he asked his helpful friend Huxley whether they would admit a representative of the Zoological Station. Huxley saw to it that there would be a positive answer. Preparations were started in a grand manner for London, but even more so for Naples, as this exhibition was to be shown first in Naples. One of the empty pavilions in the Villa Reale was rented and all of Naples was invited to have a look at the conserved exhibits which were destined to go on to London; it was enough to enter your name in a book. This meant that a visitor of official or social rank was immediately announced at the Station. Keudell was informed about the situation. »The matter has aroused a lot of attention, particularly since there is no entrance fee. I am dragging all municipal councillors there and showing them also the entire Station. All of its tables are now occupied, which proves that they will have to grant me another space.« (March 26, 1883). The following day he wrote that the most influential people of the Municipio, the aristocracy, and the press had been there. »In general there is a feeling that Naples should do something for the Zoological Station . . . they have gone so far that they are seriously considering giving me a lot immediately next to the Station. I have the impression that I will succeed in getting one.« Keudell was to visit the exhibition; Dohrn would invite the Mayor and the visit would take on a solemn character (also Virchow and the sick Secretary of State, Bötticher, happened to be in Naples. That the latter saw the lively expression of Naples' local patriotism proved to be helpful later).

Dohrn's optimistic report to the Ambassador was no self-deception. Naples had officially acknowledged the things which took place down in the Palazzo on the beach outside the public aquarium. It raised the Neapolitan self-consciousness that it was a son of the city, Salvatore Lobianco, whom even the foreigners praised as the Master, who knew how to preserve the wonderful forms of the marine fauna in such an impressive way. The City Assembly voted unanimously to present to the Institute, without cost and within the framework of the contract, a sufficiently large lot of 400 m^2 next to the Western façade.

A petition to the Secretary of Education, Baccelli, supported by a personal letter from Virchow, brought this proposal, with some hesitation concerning the amount, before the Assembly: Italy should take over the cost for the building à fonds perdu. The Department of Education alone did not have the means. But, by a lucky chance, the German Crown Prince was at that time again in Rome. In the German Embassy, Friedrich Wilhelm presented Dohrn to the present King of Italy, Umberto I. The friendship of the two royalties facilitated further attempts. They consisted – once the nobile officium had been recognized – in a more internal Italian budget procedure. The good services, for which Dohrn had always been ready, now bore fruit: the Department of Education gave 30 000 lire, the Department of Agriculture 20 000, the province of Naples 10 000 lire. At the same time, Italy's current expenses for work tables (new ones for fishery and the navy) and for subsidy of the library were enlarged from 13 000 lire to 30 000.

When Dohrn reported in the *Mittheilungen* the result of his efforts, he said that now Italy »had in a real sense given citizenship« to the Institute. And he adds the announcement: »In a few years experimental physiology and physiological chemistry will have a firm ground next to the ocean.«

In 1888 the new building was ready for use. Of course, he kept the formal rhythm created by Hildebrand. The new building served in part as a spatial enlargement of the old building, taking over the conservation department on its first floor – but after barely two decades the growing needs again outgrew the new structure.

9 RELATIONSHIP TO BISMARCK

Anton Dohrn's years of study took place at the time of the Prussian constitutional conflict. He had a personal relationship with the speaker of the most violent opposition, Johann Jacoby, and was a friend of Fr. Albert Lange, who not only formulated his philosophical Kantianism, but also confirmed his political conviction, that combined a humanistic, cosmopolitan character with a lively socio-political radicalism. In his youthful letters to Haeckel, priests and soldiers were thoroughly hated; his Jena friend received in January 1864 from Berlin the following friendly remark: »Things are pitiful here, and I invite you herewith to beat up Bismarck, together with me, on the open street.« For him on his own he found this enterprise too dangerous – the man was probably carrying a gun. That was meant as a joke; the entire correspondence took place in this tone for a while. Dohrn had announced, a little earlier, that in the case of a revolution »a saber and some roofing tile will be ready for you«. But the underlying mood is not always a joke.

The fact that it changed and how it changed, Dohrn experienced in common with a generation of the German educated class. But that process is also personal. He was not simply chasing after success. His specifically Prussian attitude was weakly developed, his strong and original German consciousness was more intellectual than political. He approved of the war of 1866 not so much as a confession to Prussia's German calling, but rather as a fight against clerical conservatism. In the campaign of 1870, the victory over the hated Napoleon III became more important than the one over the French population. National complacency, which he saw coming, made him worried rather than proud. His chosen position as a »foreign German«,* sharpened his viewpoint. He could feel among his English friends and in Italy how the increased power of the young Reich appeared from outside; not always comfortable and with identical reactions, but in a way that his pride and sense of responsibility were enlarged. Also, work in a foreign country kept Dohrn away from immediate contact with party troubles, although his lively mind followed them with affection and criticism – thus, he freed himself of certain prejudices. The fact was that, in later contacts with the representatives of Parliament, he sought and found a position above individual parties which was not only due to tactics and consideration for the Institute. Rather, it corresponded to a certain maturity which overlooked the whole. This also caused him to thoroughly change his image of Bismarck.

This was partly caused by the fact that he himself found support for his enterprise among some of the men who belonged to the closest collaborators of Bismarck: Lothar Bucher, Moritz Busch, and above all Robert von Keudell. Their representation of Bismarck corrected some of his own clichés and enriched his understanding – this happened in opposition to his brother Heinrich, who had become embittered by Bismarck's turning away from the liberals. This changed to a defiant distrust. In a letter to du Bois-Reymond (November 7, 1879) we read: »Although one may reproach Bismarck for a certain indifference to Periclean tendencies, he can still say: ›If you produce something sensible I can further each of you.‹ « It would have been amazing if Dohrn had not tried to interest »the great Barbarian in Varzin« in his attempt, with the help of Bucher and others. Moritz Busch, Bismarck's publicity assistant, was impressed by the Station and felt that it should be possible to bring Dohrn together personally with the Chancellor. »Your zoology will not particularly interest him, but the Pomeranian who was able to handle the Neapolitan devils and who could tie all the other gentle-

---

* The German word is Auslandsdeutscher, a German living permanently abroad, but retaining his German citizenship.

men to his carriage will be certain of a friendly reception.« This conversation, taking place in Capri in 1883, was reported to his sister.

Such a personal exchange never materialized. It would have been very desirable to Dohrn in those years, since he had reason to believe that Prince Bismarck was not well disposed toward him. Earlier he had noted that the Chancellor had helped him to overcome the federalist doubts of Secretary Hofmann. Not that the matter interested him very much, but rather because in 1877–78 he tried wherever he could to enlarge the competence of the Reich. He attacked the stiffening of the particularists, no matter whether they came from Prussia or Bavaria or wherever. The Zoological Station was only a minor part of these efforts, but it benefited from them.

The Naples matter may have gone through his hands repeatedly as a business problem. On May 6, 1876, Dohrn had sent his first annual report to the Chancellor, and he had written in an accompanying letter that he was well aware that he, as a German, had to work on this ground with a particular responsibility. Bismarck should please feel certain not only of his gratitude for the support of the Reich, but also of his effort to develop further his enterprise »with its European fame«. The memos of the Ambassador went to the Chancellor. The Secretary of Education, von Gossler, wrote on August 25, 1879, that one could not reproach Dohrn for lack of caution, his enterprise did »not have an equal«. He adds – perhaps in order to prevent unfriendly gossip – that, according to his source, »Dr. Dohrn and his family live in a very humble way and without any extravagance«. One could therefore count on the Chancellor's well-meaning neutrality.

Then a political incident occurred. Heinrich Dohrn lost his seat in the elections of 1878, which meant the downfall of the national liberal party. For Heinrich, who liked to breathe the parliamentary air, this was a shock. He went over to the Secessionists, the group around Rickert and Stauffenberg, who were fighting at the moment against Bismarck's political turn concerning customs, and who accused the Chancellor of unprincipled opportunism. Heinrich Dohrn did this in January, 1881, before an election assembly in Grabow. In the course of it he mentioned what Heinrich von Treitschke had said about the State Chancellor Prince von Hardenberg in his *German History*, which had been published in the previous year. Grabow was even then not the center of political events, but the speaker saw to it that this was reported in the press. Some journalists took it up and gave a sharper emphasis to the comparative points. Bismarck filed a suit for defamation of character against the speaker and the press, something that came easily to him at times when he was in a fighting mood. During the suit, the reporting journalists were sentenced to a sum of money, and Heinrich Dohrn was acquitted.

This story was probably still in the Chancellor's mind when he received the next report by Keudell in June 1882. He wrote some distrustful remarks on the margin; Keudell described the situation of the Station as »precarious«, but with a great future. He noted that »by chance« the private subsidies were somewhat diminished. Bismarck put a question mark next to »by chance«: »That will grow, the first eagerness will disappear.« When Keudell mentioned the many examples of »dangerous competition from other nations«, the word »dangerous« received a question mark. One should »not only keep the Station on its present level«, but increase it. Bismarck remarks: »How much?« Keudell's petition aimed at a subsidy of 40 000 RM, the sum he had calculated as necessary four years before.

Count Rantzau wrote on October 17 from Varzin that the Chancellor wanted a more detailed account: »He did not know what achievements the Station could show and why it had to be in rivalry with similar institutions.« He also wanted to know how the Parliament had reacted on earlier occasions. Councillor Humbert therefore had to sit down and write a substantial memo, on October 19, about the history and nature of the Station. But Bismarck was not satisfied with this overview, as his annotations on the memo show; he wanted it in greater detail: »By whom founded?« The claim of the document that the Institute »was considered a model for science« – Bismarck adorned this with the remark: »quo teste?« [who claims this?]. The frequency was supposed to be specialized. »Who was the petitioner?« follows the remark on the parliamentary initiative. And the remark that the abolition of the forced currency rate in Italy was harmful to the income of the Station he answered with the opinion that this should have the opposite effect. He also wanted details on the remark that other States had founded Stations à fonds perdu: »Where?« Rantzau demanded, on October 22, a closer presentation of the origin and wanted to know exactly by which members or factions the petition to parliament was presented. Now the extensive answer was signed by Hatzfeld, who added the vote of du Bois-Reymond, Virchow, Siebold, etc. He pointed out that the Prussian Secretary of Education described the Station »as a member – no longer expendable – in the German scientific organism (here a question mark by Bismarck). Its value had become firm in everyone's conviction and was no longer doubted.« To this Bismarck adds: »teste Virchow?« [on Virchow's testimony?].

This particular note is revealing. He searched for political connections and suspected that the hated parliamentary opponent was the special patron of this demand. He did not know that Virchow's policy of economizing, three years ago, had had the effect of keeping the subsidy below Keudell's calculation. Thus arose the somewhat paradoxical situation that the Chancellor, not trusting Virchow, unknowingly confirmed Virchow's position. He ordered Rantzau on October 30 to give a negative answer: he did not see sufficient cause for answering Keudell's petition positively. A one-time payment should take care of all pending debts. The support of foreign Stations »cannot become a reason for constantly adding so much more to the Reich's budget.« If the Parliament should show itself »particularly generous, one could talk about it once again«.

Dohrn was not surprised by this position. In September he had already written to Eisig: »Bismarck does not want to increase the budget. My name, or rather the name of by brother, does harm to the Station in the Chancellor's mind. I know that rather well . . .« The rejection at that moment was very painful. The report to the Foreign Office had given reasons for the inevitable increase of expenses: the »Johannes Müller« (the first boat) was no longer sufficient for the growing number of scholars – a second steamer had to be bought. The »competition« of the others, mainly the French, could become dangerous, unless there was, as the French had, an instrument for microtomy at everyone's disposal. Now they had to accept things as they were. In June 1884 Dohrn sent an extensive letter directly to the Chancellor, in which he tried to make clear to him his own scientific position, his achievements and plans. The Secretary of State, Bötticher, had the idea that one should, in the *Norddeutsche Allgemeine Zeitung*, start a series of good foreign reports about the Station (of which there was no lack). Perhaps this could have an effect on the Chancellor. But the atmosphere was not yet favorable enough. Keudell had renewed his petition in 1884, but had taken it back with Dohrn's approval. One must let it rest for a while. Gradually, there developed some relief for the budget, and Keudell could report a surplus in 1887–88.

Since Robert Koch had visited Dohrn in spring 1887 and planned the institution of a bacteriological department with him, the enterprise now had a better chance: the increase no longer appeared as a way out of difficulties, but as a beginning of new action. Herbert Bismarck supported it in 1889. The matter probably did not reach the Chancellor again – but certainly he no longer interfered in a negative way.

Dohrn's judgment of the Chancellor was not influenced by his personal experiences. Above all, it differed from his brother's, who fed his anger with a lot of gossip. Anton was full of irony toward him. In 1883 he wrote to his sister that he was »the all-wise parliamentarian. What is his purpose, if one does not follow him?« And also: »It has been demonstrated often enough that the man remains un-understandable until the result is what he wanted to achieve.« He also mentioned that he did not submit to any political group. The yardsticks must not be found among the commonplace people.

Dohrn took the books by Freytag and Ernst von Coburg as the unintended best justification of the man and his extravagance (1889 to his father). And in a letter to Marie Dohrn in August 1892, he says: »You judge Bismarck's speech correctly; he himself has caused much that he now blames. Altogether, he has sinned as grandly as he has succeeded. But it is necessary to acknowledge *both*; as little as I value special features, I am still glad to see all the excitement in Germany.« If these words have the coolness of the man who esteems historical phenomena as such, there grew over the years an admiration and a human warmth. As a radical young man he had once wished to »beat« the Prussian President of the Council. But now he was thinking of him all the time, particularly when, with a sharp eye and a good knowledge of the acting persons, he followed the foreign policy of the time after Bismarck. He noted more than once that now would be the occasion for great politics which could be certain of its potential. But where is there a Bismarck?

10 THE LABORATORY BOAT

It is understandable that some of his friends were a little worried about Dohrn's imaginative will to create. All his plans were compelling, but they followed each other too fast. His old friend Miklocho-Maclay, who worked in Australia, suggested the creation of a subordinate Station in Sidney. Dohrn seemed willing to accept the idea, but his English friends dissuaded him. He should concentrate on Naples – they would prefer that. In Messina, a University Station should be founded, but only if he himself would take it over. He had to proceed carefully for, although he judged the fauna favorably, he did not want to have to take an official position in the neighborhood.

While he was still working on the propaganda campaign for the *Fauna and Flora*, skillfully attributing the enlargement of the Station to a special Italian achievement, a new plan formed itself in his indefatigably working mind, one which completely occupied him for several years. He believed that his somewhat fantastic idea would find fertile ground in other imaginative minds. The time seemed to him to have arrived when the matter of zoological research should be taken out of the hands of ministries and academics, faculties and factions, and handed over to the good will of the people.

He could assume that the Station had gained some popularity, although he could not calculate its extent. In the world of science the Institute had become recognized within a few years. It was his good luck that in 1880 a teacher, Schneider, to whom he had refused a table because the man lacked the necessary scientific and personal prerequisites, had an article printed which publicized all the gossip about building expenses, about the steamboat, and scientific one-sidedness; this could only be welcome to Dohrn, who had known the man for a long time and whom, as he wrote to du Bois-Reymond, »he would throw out, even if he came with a passport from the Apostle St. Peter.« For, in an answer, a *Defense* which was extremely well formulated, witty and open, he was able to clear up the whole subject. He had the laughter on his side and put an end to the entire inaccurate gossip. That he had suffered some wrong (who were those witnesses?) confirmed the good name of the Station.

At the same time, he had a fairly public success during the Fishery Exhibition in Berlin in 1880. For a while, he toyed with the idea of appearing there as an Italian Government official; that had been discussed and would have made some impression. But without doubt the decision to concentrate on the Station was better. He had an elegant model of it made, showing fishing equipment, microscopic preparations of preserved animals, and photographs, and it made a great impression. The imperial couple visited the exhibit. Dohrn told Keudell an amusing anecdote how the courteous old Emperor spoke to the man from Italy in French, in which language he answered, to the astonishment of the Emperor: »I am a Pomeranian.« The Crown Prince and his wife treated him with distinction. Friedrich Wilhelm felt jovial as a patron and savior, and the echoes in the press were therefore very friendly. Dohrn received an honorary prize – it was of course, at that period, a drinking horn. The Station itself received a golden medal. The receiver was by now so experienced that he knew these kinds of honors were not insignificant where his larger purpose was concerned.

These things formed the background for his goal, which he had written down during a vacation in Ischia in 1879, and which he now started zealously working toward: the building of a laboratory boat. This meant neither more nor less than a floating station, but it was imagined in a way that it should serve not only the zoologists. In that period Werner von Siemens had become Dohrn's special confidant. He had once helped decisively in the Prussian Academy, and Dohrn liked to use his judgment as a touchstone, since von Siemens had recognized Dohrn's own scientific problems and his achievement, in which boldness of planning were combined with cool calculation. Dohrn presented the matter to him in November 1881, and they concluded that the building might take three to four years. By that time, the number of Station assistants would have risen from seven to ten, and successful field work would be possible. How much use the geologists and the historians would have from that boat! And the smaller islands of the Mediterranean, which were at present inaccessible, could then be reached. »Imagine making Crete the goal for two consecutive years. Crete must bring very important results to archeologists, collectors of inscriptions, people who dig. It was probably the starting point of Egyptian culture after Greece.« One could call the plan fantastic, unless the Naples Station were to represent the firm center for intellectual conquests. The old German tradition which, starting with Winckelmann,* had driven German scholarship there for more than a century, would receive enormous new

* Johann Joachim Winckelmann (1717–1768). Archeologist. »*History of the art of Antiquity*« (1764). Went from love of Roman art to love of Greek art. From that time on Homer was preferred over Virgil.

strength. During an invited visit, Dohrn had told the Crown Prince of his idea. He wanted to put him at the head of a committee, but a cautious adviser, Mr. von Normann, thought that the attitude of the financial circle would have to be clarified first. At least, Dohrn could write to his friend: »The last word of the Crown Prince was: Gain the interest of the entire nation – then the matter will be safest.«

A committee, or something similar, must therefore be formed. Would von Siemens, who was »highly interested« take the matter over? »No. Give up the idea of using me as ringleader – I do not like the boasting side of it.« But he would donate 5000 lire. What about the estimate and the working expenses? For the latter, Dohrn's imaginative mind had found a marvelous way out: he would offer his boat to the Italian navy as an exercise and school boat. Admiral Brin jumped at it. The Italians were willing to take care of the captain and the crew with all that that implied, including the coal supply. Their navy was still small. This was, of course, not a battleship, but one could train sailors on it, too. The building expenses were still uncertain, depending on size and equipment. Dohrn thought of a boat of 300–400 tons, with an engine of 150–200 horsepower.

In the *Mittheilungen* of 1884 he drew the inner organization and added a colorful and vivid description on how he imagined the economical but successful cooperation of the various disciplines. That could continue »if I wanted to stir up a kaleidoscope of scientific problems as they appear to my imagination«. What he had indicated to von Siemens by letter some years earlier was now carried further through the suggestive nature of his announcement. Negotiations were started with the dockyard in Stettin, »Vulcan«. They considered it a matter of honor to deliver the first draft of the unusual task without cost. If it should materialize, the cost would be between 450 000 and 500 000 Mk.

Being »the person who wants to replace the old with the new or to put *something* in the place of *nothing*«, Dohrn acted cautiously. He had described the situation to Eisig in the fall of 1883: »The German Reich won't give us anything (except the amount already granted), the German *people* should give.« Perhaps it was a blessing that others talked first. A friend, Professor Hans Gierke, who came from Stettin and lived in Breslau, wrote a warm and friendly pamphlet on the Station. The Crown Prince was to write a letter to the future committee – that might be useful. Dohrn did not want to write the invitation himself. »For one thing,« he wrote to Normann, »I am not skillful in writing such things, and secondly it is embarrassing to praise myself either too much or too little – both are equally dangerous.«

But there was always du Bois-Reymond, who never said »No,« if one knew how to handle his charming vanity. Dohrn sketched a letter, and Helmholtz, Virchow, von Siemens, etc. signed it; so did the conservative representative Maltzahn-Gültz; the Mayor of Berlin, von Forkenbeck; and the Secretary of State, von Bötticher – who had been won over completely by Dohrn during an extended stay in Naples. The Secretary of Education, von Gossler, suggested they ask the President of the Reichstag, von Levetzow, to invite a selected number of guests to one of the halls of the Reichstag. The latter cooperated and on June 24, 1884, this assembly took place.

As far as morale was concerned, it was a great success. The secretaries Gossler and Bötticher were there, as well as numerous scientists, some bankers, and industrialists. The friendly letter of the Crown Prince was applauded. Dohrn's programmatic speech was complemented by talks of the anatomist Waldeyer and the physiologist du Bois-Reymond. Then the Secretary of Education gave a talk. The transition of this Department from Falk to Puttkamer and then (in 1881) to Gossler had caused some worries at the time. Dohrn knew that Councillor Goeppert had supported him throughout, but he had meantime died. A new Councillor was appointed in 1882, a professor from Strasburg by the name of Althoff. The first encounter Dohrn had with him passed in an orderly fashion, but Dohrn retained a sense of uncertainty on how to assess this man. But then Gossler spoke for Dohrn with very strong personal warmth: for his humanity, his willingness to sacrifice, and his toughness. He referred to the detailed documentation and emphasized how much Germany already owed to the Institute. Central in his speech was the idea of the »Nation,« the specifically German enterprise, and the fact that someone had not waited for the government, but had dared to do what was necessary on his own. The speech was a hearty and unequivocal plea, in which the Secretary promised not only the support of his administration during his time in office, but his own willingness to take over the presidency »of a central committee for the support of the Zoological Station in Naples« – when this suggestion was made by the assembly. Local committees were expected to go along. Dohrn himself gave lectures in Hamburg, Cologne, Frankfurt, Stuttgart, etc. – a whole series of lectures was planned. But he gradually assessed his experience and found that as long as he was present in a certain place, he could fascinate the invited listeners and win everyone's good will, but he had no substitute to follow up on his idea. For the people who became members of such committees, it was a friendly gesture which carried no obligation. The president of the »Seehandlung« was asked to start the financial advertising, but did not do it. Between the optimism of Dohrn's reports to Naples (»the old battle spirit has come over me« and »in one word: great or not at all«) and the actual fund raising, there remained a gap, which did not close in the following years. Dohrn had to learn to resign himself – the »future steamer«, as he called it, was not built. The »Nation« he had wanted to attract according to the Crown Prince's call remained basically a few small circles of scientists and a few financiers, who gave him only 60 000 Mk. This sum remained apart from the Station's administrative budget. It might one day be welcome for some special purpose.

This was probably the only time that Dohrn gave up a great goal that had completely possessed him. In his report of 1892 he wrote a sort of obituary for the plan. He declared it »even today to be quite practicable and feasible«, and spoke of a »delay« for later years. »Only one thing I had not considered – how conditioned my existence was and how exclusively I would have had to devote two full years at least to carrying out the plan of the Floating Station in order to make it a reality.« »Delay« was only a euphemism for »farewell«. Later, Dohrn seemed quite happy that the plan had miscarried; it would have turned him even further away from his own scientific work. On the positive side it had brought him into close contact with Gossler – his letters to the Secretary show an almost immediate intimacy, and the Secretary was willing to take over a fourth table for Prussia.

V SECURITY AND ACCOMPLISHMENT

1 CASA DOHRN

The pompous but hazardous adventure in the Royal Castle at Portici ended in 1878. The park was no compensation for the uncomfortable task of partitioning the large, empty halls and the insufficient furnishing. The house on the Posilippo remained a dream; they found temporary quarters in the Rione Amedeo which were expensive and unsatisfactory. Dohrn was constantly busy trying to find his own house, not too far from the Station. His mother and brother advised against it – they found the general situation too uncertain. His father could not be approached at all – he had after all »disinherited« his youngest son. Frank Balfour was at their disposal and he took an intimate part in all these personal matters.

In the new part of the via Amedeo (via Crispi today), which at that time was opening up, a more practical lot was available. Balfour lent the amount. The via Amedeo stretches along the slope of Vomero at one third of its altitude. G. I. von Baranowski could produce a larger sum; the architect placed his honorarium on it, and a loan was found. In the planning Dohrn calculated that they would in the future live not only more comfortably but also more cheaply. That was not quite the case, because the 60 000 lire building costs, as originally calculated, grew unavoidably to 80 000. The builder – the ship's engineer Petersen, who for the past several years had taken care of the technical installations of the Station and who was an excellent practitioner – had made mistakes in working out Dohrn's sketches, and in his own sketch had actually forgotten the inner stairs that led from the living rooms to the dining area. He had concentrated too much on the garden and the outer stairs, which ran from the ground to the long balcony stretching all along the ocean front. The result was an organic and original building, consciously reserved, with a simplistic façade, unusually puritanical for the 1880's, but beautiful and always surprising in its connection with the garden, into whose green-shaded, carefully selected flora one steps as if into a new, fairy-like room of the house. Dohrn remained constantly anxious on his future trips to provide this garden with suitable or rare plants. It was more than a quirk that, between the street and the house which was reached only over a bridge (due to the steep slope of the lot) a bowling alley was built, leaning against the mountain beneath the street. The amusing game and the relaxed physical movement was enjoyed as a very positive contrast to the scientific research.

There stood »Casa Dohrn«, the first of a building development which later somewhat narrowed it in. In the terrible year of cholera, 1884, they moved in. The view went over the Gulf toward Capri. Green grew over the walls, marvellous palms rose high; in the unadorned, almost severe power of the construction it was nevertheless a dignified place. Here Dohrn finally found and created what he had planned in the first sketches of the Station, and what he had had to sacrifice for the overwhelming claims of the workrooms: the human center for friends and guests. In the building on the beach everything had become too narrow, every corner was used for some purpose – but here an expansive hospitality could be developed. On his round trips in Europe he had to participate in innumerable parties, often enough with secret sighs, but taking this martyrdom upon himself as a business duty. In Naples he kept himself quite consciously outside all the conventional connections which develop so easily and mar the free disposition of time and strength.

To be sure, this full life was not quite without shadows. Marie Dohrn remarked once that all people of any significance, who visited the aquarium, were expected to become interested in the Station and were treated according to their individual interests. One cannot avoid hearing a critical undertone of tiredness which resounds in her memory. In the years of unbroken freshness she achieved great heights. Her love had grown around the loyal enthusiasm and the superiority of the older man. She was unconditionally ready for sacrifices and resignations, which the »work« demanded. But this work which would be completed if not tomorrow, at least the next day, had become a tyrant. One could no longer foresee any chance of freedom. Certainly, among the scientists who worked there, there were men of high rank who went their own way. The young woman, intelligent, with clear insight, and sufficiently experienced in conventional matters, foreign as they were to her, accepted the necessities of the work. But their worth and necessity at times appeared doubtful to her. Was her only purpose, put upon her by fate, to undertake everything to master the role she had to play? Or did her inborn, growing power not seek a different goal, another fulfillment?

One can sense this question even before she is actively looking for an answer. Three little boys filled the house with their noise and a fourth boy was born in the year after they moved in – enough work, worry, joy.

Anton Dohrn felt that the young girl, at whose intellectual development he had worked from her childhood on, was developing her own features as a young woman; he watched with a mixture of pride and worry. In 1883 there had been a minor crisis. He wanted Marie to be assured of his feelings: »My love for you is not a loan which one could sell,« he wrote to his wife in 1884, »it is given à fonds perdu. I do not own any other capital. The entire talent – and it is a talent – of not feeling unhappy consists in the fact of honoring what one owns rather than looking at what one misses.«

He closed the sentence with self-irony, »Am I not very wise?« To his beloved sister Anna Wendt, however, he reported a year later, in the summer of 1885: »Marie has overcome the inner fight which she fought two years ago, i.e., to work independently, and has emerged from it more clear and noble. The perspective of our common life is shaping itself visibly more grandly, and from day to day she feels more what a great role she plays in it. She is simply indispensable in the drama.«

Precisely in those years when Italy had promised him a new building and at home a group of men came together to serve his new plans, his feeling for the general significance of the Institute grew, which had once been dedicated to an apparently limited realm of zoological research. To be sure, with the aquarium he had produced a remarkable object of general interest. Nobody of any distinction missed the occasion to visit it. It was still somewhat presumptuous when he wrote in 1876 to Fanny Lewald who, six years earlier, had been so negative: »The Crown Prince of Denmark, the King of Greece, heaps of Dukes – tomorrow I will invite Moltke . . .« But it sounds different when he wrote to his old mother in the year of her death, 1883: »Every sensible German who comes here is glad to find in the Station some patriotism of which he can be proud.« And to his sister he wrote the proud words: »I believe that nobody since Hamilton and Philipp Hackert has exhausted Naples as much as we have . . .« The remembrance of the great Italo-German landscape painter of the Goethe period expresses the consciousness that he no longer looked on himself as entrepreneur and guardian of a scientific specialty, but rather as the heir and carrier of a general tradition, also in its obligations.

This tradition received its proper form. In Florence, Adolf Hildebrand, Arnold Böcklin, Karl Hillebrand, and Isolde Kurz worked simultaneously; in Rome Marées worked hard on his last fulfillments – for these Germans the South set the tone, with the breath of antiquity, the great art of the high Italian period, the elementary features of the Southern civilization. Anton Dohrn stepped into this Italo-German tradition, to whose late romantic ending the young Carl August had once belonged. Originally he had nothing to do with all this – they did not investigate Italian history and culture. Instead, they sat on the beach because the sea was unusually rich in amazing formations of fauna and flora. There was reason for astonishment. The older scientists were sometimes angry that the younger ones did not stand in amazement, but went on to the microscope and microtome. Dohrn never lost his sense of amazement. But it was not an artistic drive that had taken him there, but the very rational desire to create an arsenal for Darwin's doctrine.

It was the effect of Dohrn's personality and also his goal that the atmosphere of the Station not only retained the character of a scientific workshop, but rather gained a cheerful and open intellectuality. That was already true when, in the first plans, he had provided for the large music room. It was true for the almost monastery-like community in the early years, high up in the Palazzo Torlonia. Now that the new house was finished, the interplay of work and life could develop its own charm. It was always a colorful, lively world, the family was constantly enlarged by house guests; the loose unconventional sociability extended to assistants and scientists, friends of the house, and travelers of rank coming through Naples. It was not a question of title and fame, but of the human level which decided whether the guest was welcome to stay and whether he himself liked to return to this »atmosphere.« The aristocratic air of the master of the house had a patriarchal flavor; but it made people feel free and secure, provided they had freedom and security in themselves. There was no jealousy or gossip, but a grand air of tolerance, which accepted and lovingly cared for the individual and knew, out of a mixture of temperaments and origins, how to create a community. Vain and egotistical people did not come; they could not tolerate the joking irony which had its home there and which Dohrn not only allowed to be directed against himself, but used as well. In this free company a sort of self-selection took place. Tensions calmed down, there was the bowling alley, billiards, music, as well as lively and rough games. These brought out the merry, childlike features in Dohrn which amazed and sometimes shocked those who knew him only as the conscientious and fastidious director of the Station. Those were also the evenings during which Marie Dohrn must have felt how indispensable she had become to the work. Above all, she helped the strangers who did not know the language and also advised them in the daily worries of life in a foreign country. Her wide knowledge of literature, her lively sensitivity for all artistic creation, and her inner participation in the great political events gave her an exalted position.

She was still young when she began this new task, but she overcame this disadvantage with intelligent, cosmopolitan security. Simultaneously, she conducted her home affairs with circumspection and a permanent readiness – for from the Station there always came, by accident or necessity, the demand for improvised hospitality.

She had »Augu« on her side, a Swiss-German lady from Catania, well versed in languages, who had suddenly been thrown by fate out of her bourgeois, pampered security. She entered Dohrn's house to take care of the children and remained there decade after decade. One could not imagine the Dohrn household without this quiet and active participant in all of the family affairs down to its grandchildren. Beginning as a servant, she later grew into a powerful and good mistress. One cannot imagine the inner history of the »Casa«, mainly in Dohrn's last decade, without her – she was a beloved and honored person who demanded respect.

Those evenings when music and good conversation and social free-for-all came together became for many participants a happy memory. They belonged to the nature of the »Station«, just like the trips on the »Vaporetto«. This steamboat did its service trips with fishing and dredging; but it also had the function of taking people to beautiful places in that incomparable landscape.

Dohrn was sometimes reproached by certain foolish people for seducing people with his little steamboat. He laughed at it – work always came first. But since this story of the pleasant steamer was gossiped about to some extent in Europe, there were sometimes unpleasant and painful stories, as when a stranger was of the opinion that a boat trip was part of a visit to the Aquarium. They had to make it clear even to German princes that this was not the case. But Dohrn fully realized that the little adventures and surprising beauties of such trips were the most pleasant advertisement. They were also, for the co-workers, a form of relaxation in the hot days of that enervating climate.

The »Vaporetto« carried the flag of the German navy. Dohrn announced once that he wanted the boat to carry it, and then he did it without trying to find out whether that was correct (the Steamer was listed in the Stettin boat register only in the 1890's, when the German consul thought that, because of the Ocean Union, the boat ought to be registered somewhere). Dohrn realized that the appearance of a German flag, particularly in the small coastal towns which were not visited by larger boats, brought joy. He considered this a nice addition to his free mission as a German.

2 THE RISE TO FAME

The system of work tables, with their international character, was the main foundation not only for the current finances, but also for the particular position of the Naples Station. It was not, however, a particularly secure foundation, because it depended on changing political budget situations in the individual countries. Already in his first letter to the Secretary of Education, Falk, Dohrn had said that he would not and could not give the work tables to »private persons«; it should not become a »privilege of the rich«. Only in a few cases, when tables were free and foreign scientists needed the Naples Institute for special investigations, did he make exceptions. But the rent did not mean security. Taking over the tables was in no way done in a unified kind of contract concerning its duration; they were for one year, two years, five years, and the renewal was not always assured. At the moment when other nations were following Dohrn's example, but with government support, new uncertainties arose. Austria disappeared for a few years, later in Russia it repeatedly became a matter of power: the Secretary of Education considered the Naples money unnecessary, since Russia owned its own institutes on the Black and White Seas.

Although Dohrn had gained a very good position in St. Petersburg, mainly with Secretary von Reutern, critical situations arose nevertheless, which were partly under the political influence of

Paris. But the Russian scientists who had worked in Naples spoke through the universities in favor of the continuation of the Russian commitment. That was an important event and a good argument, when here and there the ministerial bureaucracy became eager to cut out their contracts, as, for example, Prussia in the mid-1890's. With such an attitude among the scientists, it became clear that Dohrn could always feel that he was the giving person. Mainly the older generation felt this – those who had worked on the lower marine fauna before the Naples Station existed and who therefore knew from their own experience what it had offered to every biologist.

In the first years Dohrn had to »acquire« – he never quite stopped generating and expressing wishes. Later the judgment of scientists, the quality of their research in Naples, the name of the Institute served as advertisement. In 1880 Belgium joined the circle of renting States; in 1884 it was Romania. In 1883 Dohrn had visited Bucharest and was met with lively understanding by the Prime Minister, Stourdza. For several days he was a guest of the Royal couple and was much impressed by the humane manner of Carmen Sylva. Normally he liked to add an ironic tone to his letters in which he reported meetings with crowned heads. But in the extensive description he sent to his sister, this tone is lacking. And yet the queen-poetess treated him at length with readings from her work. But he gratefully felt towards her only warmth and kindness.

At this time he also tried again to find connections with France. Dohrn did not initiate this – he had not forgotten Carl Vogt's words in 1873, mocking any opinion that the French State would take part in an Institute created by a Prussian. He was professionally eager to find a contact in France. Some Stations had been founded there which promised good work, although they belonged to universities and their goals were more limited. He saw in Lacaze-Duthiers, who worked for the same goals, an important competitor. There was no lack of personal relationships, which achieved the character of intimacy. In tolerant circles of French scientists the demand came up to participate in the continuous exchange of experiences and goals which took place among the biologists in Naples. Secretary Ferry, who had hoped for a while to clear up the German-French relationship politically, asked for professional opinions on the problem. But he made no progress, for it was Lacaze-Duthiers who most strongly opposed it. Dohrn reported to his sister in 1882 that »the famous French zoologist, my special competitor, answered: If the government takes a single table at the Zoological Station of the Prussians in Naples, I will destroy my Institute in Roscoff, I will take my leave as professor at the Sorbonne, and return my medals.«

It was not worth the Secretary's time to carry on the conflict, and France remained out of contact. Access was given to a few scientists. With the Parisian physiologist E. J. Marey, for example, a good friendship persisted. In 1896 Dohrn was his guest for some days and was most interested in Marey's attempts to make the recent technique of »Cinéma« useful to scientific knowledge. F. M. Giard of the Sorbonne tried incessantly to initiate permanent contact. His successor, M. M. Caullery, wrote as late as 1906 in the *Revue Scientifique* quite emotionally on »why France should have a table – almost all foreigners who count have always been there several times.« France, he said, had opened too many Stations through its universities, all of which suffered from miserable budgets, and even the best endowed did not equal even a fourth of Naples. It was the French who would suffer most if they kept away from this »permanent congress«, in memory of the time of its foundation. The voice of a preacher was sounding in the wilderness.

It was important for Dohrn to win over America. The relationship with Dom Pedro II of Brazil, who had a lively interest, remained in the realm of anecdotes. In the United States it was difficult to find a beginning. There, the two Agassiz, father and son, had given a strong impulse to the investigation of the marine fauna, but they tried to monopolize any possible money offered for their own research. The financial question was not so much in the foreground of Dohrn's wishes as the scientific necessity of including their young movement in the problems of European science. A coincidence helped. A wealthy American, Major Alexander Henry Davis, who was later to become a good friend of Dohrn, offered to rent a table for American scientists at the end of the 1880's, the so-called Davis table. Later, the Smithsonian Institution in Washington took it over. Particularly among the young American scientists some, such as E. B. Wilson, became close to Dohrn on a human level and were full of gratitude. Finally, American participation in Naples grew to five tables, while some American Institutions remained poor and Dohrn offered to intervene with the American Ambassador in Berlin. Two of these tables were rented by the Carnegie Foundation, one by the Smithsonian Institution, one by Columbia University and one for women scientists. A special association for women had been founded over there.

It was a satisfaction for Dohrn that in Great Britain, although several Stations – above all one in Plymouth – were founded, Oxford followed the example of Cambridge. Ray Lankester, who had worked with Dohrn already in 1871, had become full professor and had gradually overcome the resistance of traditionalism: Oxford, too, took a table. In connection with the Italian financial situation and, considering the constantly perfected equipment, Dohrn petitioned in 1883 for an increase of the yearly table rent from 2000 to 2500 lire. This caused no trouble.

In the first years after the foundation of the *Mittheilungen* Dohrn had added to the scientific contributions an overview of the development of the Station, enumerating the scientists who worked there and cataloging all their publications. The overview of the years 1885–92 is the last of these statistics. Later, the regular official reports were enough; a regular format had evolved for them. In that Report, the number of scientists had already reached 709. These were names that became great and internationally famous, but also some that remained a mere statistical note. Among them was a young Norwegian. Although the Scandinavian States had no tables, the 25-year-old Fridtjof Nansen[30] had written from Bergen to ask Dohrn whether he would give him the opportunity to work there; he needed this for his future plans. He was accepted. Nansen stayed in Naples for several months in 1886 and later reported in the Bergen Journal *Naturen* about the kind of »scientific stock exchange« – this was his impression of the Station. He prepared plans for a Norwegian Institute, but left it to others on account of his trip to Greenland. The young, slender Norwegian was considered the merry heart of the society. Dohrn himself noted: »He had nothing against the pleasures of life and danced well.« Later, everyone enjoyed the memory of having met a young man who faced his future, free and secure.

Since March of 1879 Dohrn had held the title of professor. In the beginning he had fought a little against it; he preferred to remain Anton Dohrn and disliked university life, thinking for a while of Jena. But the kindly Goeppert of the Prussian Ministry of Education advised him that the title would become important for the planned parliamentary action. Dohrn understood. Altogether he had to see how his position was growing into an official form which he had never contemplated, but which he learned to master skilfully. He had already experienced the meaning of a medal in 1875, and he received another one, the Mauritius, with which the Italian King endowed him in 1880. A Russian medal and several German ones followed. In 1890 the Prussian Secretary of Education in Berlin noticed that a Prussian one was missing – and yet the Hohenzollerns had shown their friendship to the Institute. That made for a nice correspondence. The third grade of the Red Eagle was suggested, but Althoff wondered whether this was useful for Dohrn and »his position and work in Italy«, and he asked him »for help« in the matter – he might also choose the title »Geheimrat«.* On April 14, 1890, Dohrn answered: »Since I already own two decorations in my buttonhole, four around my neck and even a Star on my chest, I would like to suggest the title of Geheimrat. This will give my position in Italy the kind of support which is connected with such honors in this country, which is so sensitive to such

* Privy Councillor. The highest German title for intellectual professions.

things. It will also serve me well in Germany, whether it is in society or in contact with scientists, young and old, who are sometimes rather difficult to tame.« Since he had not gotten the M. D., as Althoff surmised, the title »Geheimer Regierungsrat« might be suggested. »I have some practice at the Station in governing rather obstinate people.«

Gossler had started this business with the decoration because Dohrn had asked him whether his oldest collaborators could receive the title »professor.« »Many younger scientists who had been assistents only a short time are now full professors.« It would be good if the older colleagues whose scientific achievements are recognized (the Ministry demanded professional opinions) could also receive »the personal aura«.

This is only a small incident. It belongs to the image of Dohrn the diplomat who had tenaciously insisted that the architect Profumo and the navy lieutenant Chierchia should receive Prussian orders. He also suggested to the English, who had some honors in mind, to distinguish Eisig, for whom this would be a greater satisfaction. This question of a correct »dosage«, which he finds important for his Italian helpers and his permanent collaborators, occurs in his professional correspondence: His »Parsifal period« was finished, and Dohrn became a technician of human psychology. Of course, in the end he did not escape the Prussian Order. The talks between the Foreign Office and the Ministry of Education about the kind and class of the Order seem almost rather amusing. It was hard to place Dohrn into any category; he had no position, no years of service – finally, in 1893, he received the »Crown Order, 2nd class,« which the Emperor himself put into his hand.

One episode in his life was the problem of whether he could and would like to return to regular academic work. While he was a schoolboy he had considered this to be the goal of his life. Carl Vogt had suggested it in 1877, when the difficulties were great. He should gain a Naples professorship and influence the Station scientifically from there, but he should give up possession and administrative responsibilities. This friendly way out no longer needed to be taken. In 1897, the University of Naples offered him the position, which had become free. They felt that the Institute was by now firmly grounded and would not need him so much any more. Those who believed this did not know that he was already thinking of doubling the Station. He thanked them.

More exciting was the question which du Bois-Reymond asked him, first in writing and later in person in 1883, whether he would be willing to take over the professorship and direction of the Zoological Institute in Berlin. For this Institute a large new building was being erected in the Invalidenstrasse; and Secretary von Gossler seemed to like the idea of an offer. Obviously, the idea in Berlin was that the work there, which had somehow dried up, needed a powerful man with fresh initiative. Dohrn's answer is typical. He thanked them for the request because it could be useful to him in negotiations about a new building he had just conducted with Rome.

He suggested to du Bois-Reymond that they should separate the work as professor from the directorship of the Museum. They should be completely separated and the Museum should obtain its own research – for this position Möbius in Kiel would be a good choice. He himself might be willing to take over the re-organization for five years if he could spend the summers in Berlin and the winters in Naples – he felt up to the twofold task. Such a lengthy but not permanent assignment was not acceptable in Berlin. The re-organization of professorship and museum direction, and the division between the collection for visitors and scientific material was taken up, according to Dohrn's suggestions, and was one of the first decisions by Althoff.

3 SEPARATIONS AND ENCOUNTERS

The support Dohrn found in the official German offices in Italy had for a while been a question of life or death for his enterprise.

Robert von Keudell, angry that he had been left out of the negotiations about the prolonged »alliance of three«, left the Palazzo Caffarelli in Rome in the summer of 1887. His farewell letter to Naples contained the sentence: »Nothing in Italy has been as close to my heart as your work, including the landscape and the personal background.« The relationship had reached a natural intimacy, despite the difference in their ages. The ambassador was not interested in the scientific details. More than once he held back Dohrn's desire to »confess« because the word Darwinism appeared to him somewhat objectionable to Parliament and officialdom. Dohrn, being warned, tried to clarify these matters. He did not want to be taken for Haeckel's kind of Darwinian, which was viewed in those circles as demagogy. He had, by contrast, taken over the task of statesman. Certainly, he did not want to interfere, he »could not let the fulfilment of one's life be minimized by such considerations« (1882). There is something moving in the fact that, even years later, Keudell advised his friend in official communications to emphasize his own works »yet without mentioning the name of the great Englishman.« Dohrn did not follow this advice, but he knew the attitude which caused such warnings. In the next decade Keudell became something like a voluntary ambassador of the Station in Berlin.

This was a great help for Dohrn, since Werner von Siemens had died in 1892. Siemens had always lent his authority to the Naples establishment and to its leader. Their relationship, born out of Siemens' professional interest, had achieved a very personal aspect: Dohrn supported Siemens' work on his autobiography with his friendship and help: »The world, our world, has become a great deal poorer since he died. I shed hot tears.« In an obituary in the *Mittheilungen,* he thanked him for the »most profound importance« the dead man had had for the Institute. In the same place, he also spoke warmly of the General Consul Otto Beer, who had died in 1891 – his advice had accompanied the financial and organizational development. He was co-owner of a German bank, an independent industrialist, and had more than once been a support. Dohrn speaks in this obituary of the fact that Beer had given him large sums as credit because, as a financial official, he saw how skilfully and accurately Dohrn carried out his administration.

Now Naples received a professional Consul, von Rekowski. Things started well, since the new man was much impressed with the achievement he saw before him, and Dohrn, with the great experience he had gained, could become his guide. Once in a while the relationship cooled off – the General Consul was somewhat sensitive when he realized to what extent Dohrn had become the center of official and courtly matters; but this never became serious. For every official demand Rekowski was a very loyal and conscientious supporter.

Rome, however, was more important. It must have been clear to Dohrn that fate would not immediately send him another German Ambassador like the previous one. Keudell, who was not on good terms with Herbert Bismarck, and must have experienced Holstein's secret activity against the former confidants of the Chancellor, had already for some years been forced into political resignation. Under these circumstances, his active participation in Dohrn's work had been very satisfactory. But would not his successors feel this to be a tedious heritage that they were constantly burdened with Keudell's helpfulness concerning something that was foreign to them? Dohrn had to find out how to handle this diplomatic game, which was not always simple. He succeeded with Keudell's successor, Count Solms. It would be good to get him to come to Naples. »Then I will go with him wherever he wants and present this as something of a medical duty. Once I have him, he will certainly serve me.« This he wrote to Eisig in a strategic plan in the fall of 1887. He succeeded as he reported in 1889, and Solms thoroughly reviewed the Station and was much »nicer«. They went on trips on the Vaporetto and became better acquainted. Dohrn even overlooked the somewhat self-centered vanity of the man.

Bernhard von Bülow was no problem – Dohrn's openness to the world and Bülow's eager pleasure in unusual things intersected. His successors rarely rose above a coolly correct behavior. Of the inoffensive Graf Saurma a characteristic anecdote was recorded:

Dohrn wanted to talk with him of Bismarck's *Gedanken und Erinnerungen*. They had just been published and moved him very much. Saurma answered: »He had not yet read them, he had also been present at the events – so they are not so foreign to him!! That is also a point of view.« It becomes more difficult with Count Monts. He was not so old, but, together with a critical intellect, rather indolent and over-sensitive. A bad encounter occurred, which later became a comfortable and richly contrasting story, but was painful at that moment. The visit of the Emperor Wilhelm II was announced – the Emperor always visited the Station on his trips to Naples. Dohrn went to the Station to receive him. The Ambassador had already planted himself there in order to be ready, and he made a scene because Dohrn appeared in a soft hat. They sent for a top hat instead and it came in time, but then the Emperor arrived and exclaimed: »Well, you in a top hat! For your sake, I dressed in civilian clothes!« Dohrn answered with a gesture to Monts: »On higher orders, Your Majesty.« The waiting journalists saw the intermezzo and the Naples press discussed with gay seriousness whether Dohrn might wear a soft hat in front of his Emperor. Dohrn wrote to the press, »He may.« Later, he often laughed about this event, but at the time he felt, in Monts' brusque way, the attempt to humiliate or the expression of courtly inferiority. For a long time the Amabassador still resented his light-hearted reaction.

Dohrn could say, »He may.« With the death of Emperor Friedrich III one of his early supporters was lost. Together with Keudell, he sincerely mourned him. Keudell, whom he had asked to open the way to the young Emperor for him, thought he could not help him much, because his acquaintance was too distant. But when a first possible trip of the Emperor was considered, he wrote to his friend thinking about the latter's manner: »If His Majesty is really coming, I recommend, even if it were 8 a.m., to wear tails, white tie and decoration – that takes the place of a uniform in the eyes of a gentleman used to the military.« Such advice was not unnecessary. Dohrn was not informal. He had the elements of natural dignity, but it was natural; and in Italy he moved within an etiquette whose frame and limits he himself decided. This was true and, as it turned out later, proved successful also for his relationship to the new Emperor.

In the fall of 1888, as Wilhelm II took his first official visit to Italy, Dohrn waited for him in vain – the very long Navy parade overthrew all the plans. In 1891, Gossler had arranged an audience in Berlin. In April 1893 the Emperor and his wife visited the Station and were highly delighted with what they saw. »There was almost a revolt of the entire retinue, to get the Emperor out of the Aquarium. You can imagine how little I did to help them,« Dohrn wrote to Keudell, »I also have the certain impression that in all this there was as much personal interest involved as I could wish in order to be certain that I can count on the element of personal good feelings, if circumstances should ever force me to appeal to almighty intervention.« Dohrn was not wrong in this feeling.

Wilhelm II repeatedly visited the Zoological Station. Once in the Spring of 1905 he took along to the Institute the Italian King Vittorio Emanuele. In his *Denkwürdigkeiten* [Memorabilities] Bülow related with an ironic undertone, that during his second visit in 1896 the Emperor informed his retinue »in a lengthy lecture on the doctrine of life, particularly on zoology,« which filled them with admiration. Dohrn, who sometimes was perplexed by the political speeches of the monarch may not have disliked *this* speech. Whether correct or incorrect, did not matter. The Emperor saw what was being achieved and investigated – he also had a feeling for the personal charm of this man who knew his way with people and the world – Dohrn knew how to use this intelligently and tactfully. The Emperor had become important for him; in decisive moments he could handle him.

But aside from this, Wilhelm II constantly occupied his mind as a figure of German destiny. He was ready to allow him not only his gratitude, but also human sympathy. To his close friend of Jena years, the wife of Ernst Abbe, he wrote a few sentences in 1896: »We have become good personal friends; and if I still consider the young gentlemen a rather dangerous firework on the throne of the German Reich, I cannot deny that I like him personally and feel decidedly that I would like to continue the acquaintance if there is a chance. It is clear that one cannot fight with a Majesty – as in old, or even in recent times between Ernst Abbe and Anton Dohrn – but you know my imperturbability and you can trust that I am not blinded by the radiance of the Imperial dignity. I can comfortably set opinion against opinion and bring about some amusing discussions. I am sure both of us think better of each other than we did at first, although he thought better of me earlier than I did of him.«

In the few years when Dohrn kept a diary, 1895–99, much is said about Wilhelm II: opinions about him from the conversations with the upper bureaucracy, with courtiers, parliamentarians, also anecdotes which were passed on. There is great uncertainty in trying to form one image from the contradictory features of the Emperor. Dohrn also described the many encounters in Naples, in the castle of the Hohenzollerns, in Berlin. It is typical that on one occasion someone in the Emperor's environment asked Dohrn not to give up contradicting whenever he thought it suitable. Since he was persona grata, he could allow himself such things; and since the Emperor was not used to this, a clear attitude toward him was educational and useful. The Emperor trusted him. He saw before him a man who knew Europe and could judge the political leaders of many States. That was rare in Germany. Therefore Wilhelm II talked openly and without restraint to Dohrn, who honored this as proof of the Emperor's confidence, although its content worried him. But his loyal discretion was complete when it was a matter of high politics. Dohrn jots down, for example, on November 19, 1898, as the Emperor was returning from Palestine, »the strangest conversation«, during which »the most unbelievable things« were told, almost all about foreign policy. The reader has become curious after this strong description. Then he goes on: »I pass over the entire conversation, since it obviously presupposes that I, as a reliable man, would not misuse the situation and could be trusted.« And then again, »I'll keep this, too, to myself.« The reader thinks:

What a pity! But such slight remarks are psychologically typical; the critical independence and complete inner freedom filling those sheets stops as soon as he becomes reflective. A break of confidence through some accident that might happen to the papers might have unexpected consequences for foreign policy.

The worry about the Imperial policies remained. It had been enlivened by a wire sent to the President of the Boers, Paul Krüger.* If this announcement revealed a general purpose, a clarification of power was first necessary, and then – a Bismarck. Dohrn was deeply impressed by the confusion of moods and judgments he had met in the beginning of 1896, both in official and parliamentary circles. Where could the man be found who would lead Germany out of these many unclear situations?

The person who trusted himself to do so had lived since 1893 in Dohrn's neighborhood. The personal relationship with Bernhard von Bülow** had easily grown into an informal intimacy. The father of the Ambassador had groaned when the »Dohrn-infected« Keudell brought a zoologist to the office, but the »relationship« with the Station for which he was ready, had achieved the respectable traits of serious permanence. But what a change had occurred since then in Dohrn's position! In the 1870's he had needed recommendations and introductions, particularly among the Italian offices. Now, he could almost grant them himself, but with tactful intelligence he refrained from any interference in Italian matters. There was probably no other foreigner who knew equally well former, present, and future Secretaries, who had friends in the government offices, the Senate, the Chamber – perhaps also some indifferent or opposing people – but he knew them. At the universities there were scientists who owed gratitude to the

* A wire of congratulations sent in 1896 by Wilhelm II to the Boer Paulus Krüger, President of the South-Afrikan Republic. It caused hostility between England and Germany.

** Bernhard von Bülow (1849–1929). Ambassador in Rome, 1893–1897. Then Secretary of State, later Chancellor of the German Reich.

Station for science and, what was even more important to their growing self-awareness, relations to the world. Dohrn no longer needed the support of the Roman parliament – his personality had become an independent power in that country. An intelligent ambassador understood what the Station and its pioneering achievement could mean for himself. That is what Bülow did. Besides, who passed through the house on the beach year after year? English, Russian, and American intelligentsia, the Spaniards, Dutch, Romanians – they all came. This »permanent congress« was certainly not a political group, but one could hear, from its voices, what moved the educated classes of their nations.

Dohrn's notes indicate how freely the two men talked to each other about »the unavoidable topic of His Majesty« and the problems of high politics. (»We agree that Trentino should become Italian, the sooner the better, in order to prevent a continuation of the Italianization. Then the Germans and Italians can confront the Slaves.« June 13, 1897.) Bülow »emphasized (in that conversation) that his father was a very educated man on the basis of Goethe – that explains the basis of *our* sympathy. Our roots are in the same foundation of education, we know the world and man, and do not worry about appearance, etiquette, philistinism, etc.« He accepted the word »congenial«, which the Ambassador used to describe their characters.

After four years, Bülow left Rome and took over the office of Secretary of State in Berlin, as a step on the way to the office of Chancellor. This was a personal loss for Dohrn, namely the opportunity to participate in stimulating, irresponsible conversations full of anecdotes which both men knew how to cultivate and enjoy. For the Station it was a gain, that its founder knew a man at the top of government to whom he had quick and intimate access. In the very years during which Bülow was in Rome, the Station had seen its greatest development and effect. Not only the personal sympathy, but this impression accompanied Bülow to his new office.

4 MARIE DOHRN

As Marie Dohrn grew older, she retained the image of confident mastery; on meeting her, the impression was of strong humanity, which in her rich, full life had not grown weary, but remained active, participating, and adventurous. This was evident in spite of a certain harshness. Her features could not be called beautiful, but they were shaped by her strong intellect. One could still feel in them the charm of her youth. Slender, with a delicate bone structure, she was, while she was in good health, full of never-tiring activity, a bold rider, and a lover of daring enterprises, even if they were dangerous. Her oldest son remembered a trip by rowing boat from Ischia to Bajae, which was very long and stormy – her will forced the over-daring enterprise – she wanted to be back on the mainland, in order to receive her twin brother Casimir, who had come there to die. The immediacy of her nature also caused her quick and unhampered lightness in talking to the Italian peasants and fishermen. She spoke and understood their dialects and enjoyed great popularity among them, also because in the times of cholera she was helpful and unconcerned. On the Tuscan coast, in Forte dei Marmi, where she later often stayed, the love of the simple people outlasted her death.

In the early 1880's her marvellous freshness was clouded by severe sickness, which for two decades never fully passed, but returned again and again, with months of dangerous illness. This meant a new worry in Anton Dohrn's life. How much it weighed on him is clear from his letters to Abbe. What shows in these letters about Abbe himself makes the reader simultaneously admire him and fear for him because they indicate an unhealthy sense of self-destruction: he was over-worked, his nerves were bad, he took stimulants and was fanatically concentrated on his work. Dohrn warned him and asked him to understand his limits. »I am hiding behind the same shrub.« He wants to lighten his friend's burden by describing the fatefulness of natures of their kind: the burden of a responsible life, the eternal fights against stupidity, laziness, and meanness. »In one word,« he wrote in a melancholic letter of May 9, 1888, »we have become old.« He continues: »What hastened this process in my case and broke in upon me almost overnight, is the conviction that my wife's health is almost destroyed, that she is no longer up to the climate, nor to the claims of our exciting and tiring life, that she can no longer be either my old, ever-ready companion, or the secure mother of our boys. Together with the feeling that after a fight of 20 years I can still be shaken down to my roots by some silly wind – that has suddenly aged me by more than ten years. It has changed the former hothead, although not into a pessimist, but only too often into a grumpy man somewhat tired of life.«

This is the language of depression. It is not frequent in Dohrn's letters; he knew the situation and could remain silent. When the cloud passed, a certain talkative optimism started – this one can clearly feel. To Charles Grant, who understood him so well, he could speak in dark tones. But this English friend was dying. Now the confidant of his Jena years became the companion of his dark hours – he would know how to weigh his words.

What one can see in these lines is not only the sorrow of a passing exhaustion, but rather the fact that Dohrn is no longer quite sure of his own physical health. The climate is harming him as well – they grew used to taking short vacations on Ischia or finding refreshment in the mountains. But how long would this last? He wrote to his friend that he had to separate from his oldest son, who was sent to a German school – when would the second and third follow him? »And then, father and mother will follow, together with the fourth, and I shall try to fulfill my Naples obligations from Germany – that should no longer be so difficult – I hope.«

The sons should go through German schools, tutoring at home was only a poor solution. But the separation was very hard for Dohrn, who as a father gave sensitive attention to the sons' individualities, which appeared early. Hence the plan, which could not be carried out, to transfer the weight of his life perhaps to the

North. For at least the summer period he never quite gave it up. Dohrn was so much tied to his home land that he calculated whether there would be enough money to build a summer house in the Buchheide* near Stettin – forest, walks, coolness, solitude. There were times when he was almost afraid of the enervating heat and the unyielding light of Naples. For tourists this was a different matter; he, in contrast, stood in the midst of tense work and eternal responsibilities. At such moments, when he thought about his health and the fate of separating the family, Goethe's word from the *Elective Affinities* appeared to Dohrn, who thought it was meant for him: »No one walks with impunity under palms.«

In the early summer of 1888, Marie Dohrn went to Dr. Kussmaul in Heidelberg, the »wisest clinician,« to find out »whether he could help her out of the morass of her gastric-anemic situation to healthy nourishment, i.e., healthy nerves.« These are Dohrn's words. The doctor did not succeed. The sick woman agreed with pleasure to the doctor's suggestion of a long stay in the Black Forest. Later Starnberg** was chosen, where their Florentine friends, the Hildebrands had settled in the early 1890's. The regard for her health was not the only cause of their separation. It lasted for years, often interrupted by longer periods together in Naples, in other places, at the houses of friends, at family reunions, but it was a shadow on Dohrn's life.

The development of the relations between husband and wife can be touched upon only lightly. In the early period of the marriage and the common work the personal achievement of Marie Dohrn can barely be praised enough. On it rested the beauty of the endeavor and of a fulfilled happiness, despite sicknesses and worries. When Anton declared her, in the »drama« of his life, absolutely indispensable, she was for him not only the wife, the careful housewife, the mother of his children, but also the indefatigable helper

* A country town near Stettin.

** A pretty town south of Munich.

in his innumerable writings, the counsellor, the reliable assistant in social representations. She was also the patient, untiring, strong help in the bad times of depression. Coming from a family tradition of the Russian-Polish aristocracy with a sense of authority that she never lost, she had tied herself not only to the beloved husband, but through him to the creation of a bourgeois-scientific business for which she was not personally suited. The unusualness, boldness and breaking with all conventions of the enterprise had lent her superhuman strength, but when Dohrn had finished his task, a certain weariness fell upon her.

It came at the moment when a relief in the work occurred. Since 1883, Hermann Linden had taken over the direction of the business side. Now Marie could devote herself more to the children. But she had to separate from them – she saw this in her very clear mind; she did not for a moment fight against the fact that her sons obviously should grow up in German education and culture. But, being still so young, she could work out how many years were to come when all the sons would be gone from home, the house would be empty, her duties would be less – an elementary impulse rose in her against this. She did not want to be separated from the youngest, and wished to remain at least close to the older ones. The children might now need her more than her husband. The fighting man had been closer to her than the victor, who nevertheless could not give up his continued planning. Had it not been her belief and his promise that the time of maturity and rest would be a common one? He himself is sometimes resigned to early aging. »I know only too well,« he wrote in 1889 to a former colleague, »where the strength went that I don't possess anymore.« But something occurred whose effect could no longer be extinguished and was to continue beyond the individual life. »When I almost daily realize the fact that my creative period is over, I am comforted by this.« He felt that the time of his creativity was over, although the scientific problems still made him restless. But Marie Dohrn, 16 years younger, still had to assert and shape herself. A pride of inner freedom, not without harshness and bitterness, appeared in her, and the relationship between the two entered the state of critical, even tragic reflection.

Anton Dohrn himself, arguing in his own manner, analyzed the crisis in 1894, which in 1891/2 began to dominate his consciousness: »The Polish woman is revolting against the German – this was the start of your separation from me. In the same way your rebellion began against the man of science, originally caused by the pressure of Grant's antipathy against everything that science really was, and which later developed in him instinctively into a rivalry against me, of which he was not conscious. V. . . continued this, while both were kept afloat by the scientist, whose true character nonetheless was that of an artist. After the ›Polish woman‹, you began to be dominated by the doctrine of the freedom of women . . .« In these sentences there are words typical of his self-observation. In an elementary way he feels himself to be an artist, but he had written to Marie already in 1892 that »it was the moral man in me who overpowered the esthetic one.«

In his youth, meeting George Eliot and becoming friends with Fanny Lewald, he felt them to be only unproblematic representatives of feminine independence. But now this worried him. He noted in 1897 that, when the »feminist question« was brought up, he answered with a harsh rejection. And when he met the mother of a young zoologist in Munich, he said: »It is very refreshing to meet a woman who finds complete satisfaction in the ideals of the past and who nevertheless towers, intellectually and morally, over all the torn, insecure modern women.« This is not the expression of a narrow-minded sense of well-being, but the harmony of a family resting in itself, in the relationship between parents and children and among the siblings. He said that he would like it to be more amicable, freer, and more open than he had experienced in his own youth. On an elementary level he could be certain that Marie's open-mindedness agreed with him, but the immediate relationship remained open to changes. There were times of quietness and understanding, others of distrust and torment. He wrote: »I received a good letter.« Or he put a letter aside, afraid it might disturb him. This is the conflict of two significant human beings with sensitive nerves, who attempt, finally, to overcome their »misunderstandings« and find again the happiness of the ten to fifteen years of their common beginning.

To prove his willingness to make sacrifices, Dohrn took part, in the winter of 1893, in an enterprise which, although enlarging his experience in life, cost him much time and inner strength. We hear about this drama, in which he played an unexpected role, in a letter to Abbe: »A marriage conflict of my father-in-law's brother (where in Russia are there not such conflicts? They are the norm!) caused a violent decision of the so-called justice: on the basis of a false denunciation by his wife, my relative lost his parental prerogatives. His five children were taken out of the house by the police and given to his wife. The injured man defended himself and my father-in-law took part in the war. It became so threatening that I finally, to please my wife, began to take an active part . . .« This became a strenuous and upsetting adventure. »I am paying this price,« he justified himself to Keudell, »for the happiness of sacrificing myself completely to my old father-in-law and my wife by helping them to find their peace of mind.«

He worked for this matter for three and a half months in St. Petersburg and Moscow. It was a campaign in the atmosphere of the Czarist court. He had already traveled back and forth in Germany. At the Court in Darmstadt he interested the Grand Duke Sergius in the problem; they contacted the son of the Chancellor, Alexander Hohenlohe: would it be possible to reach the wife of the Czar by means of her sister, Princess Heinrich, who was in Naples so often?

On Russian grounds there was a very complicated strategic situation. Georg and Alexander Baranowski may have been happy that such a reckless helper appeared on the scene, but Dohrn believed that the two men, both belonging to »high society«, had started everything unskilfully and plainly wrong. He had to push them somewhat into the background, which was not feasible without annoyance and mishap. The »bulletins« which went almost daily to Marie Dohrn are written in a colorful style: if he reaches one Secretary at one place, at another an Adjutant-General escapes him. He must make it clear to everyone that an overpowering of Justice and also a disregard of the formal judicial process have taken place – how does a German professor get mixed up in an internal Russian affair? He would like to ask this of himself: the presuppositions under which he started the matter were not to his liking, the environment of *this* Russian upper crust he found disgusting. But he was seized by a sort of sporting ambition not to leave Russia before the matter was settled, even if he had to go as far as the Czar himself. The German diplomatic representatives could, of course, not be of any use. For them the eagerness with which he took care of this family atmosphere of the Russian nobility was rather inconvenient. Dohrn did not get beyond the antechambers of the Czar, but he finally left there as the victor: The uncle of Marie Dohrn got his children back.

The result of his enterprise did not quite correspond to the effort. Some personal injury remained – aside from the satisfaction – that the *German* had helped where their own efforts had failed. Also, Dohrn gave in his letters and conversations a very drastic expression to his impression of the Russian-Polish judicial system and moral condition. Without his wanting to do so, this left some scars on his wife. She had a certain affection for both her father's and her mother's heritage. Also, there was in White Russia the large estate, Wydranka, which she felt should be kept for one of her sons. When Anton Dohrn thought about Wydranka, he felt uncomfortable. It occupied the imagination of his wife more than he liked. Keudell, who clearly saw this antagonism of interests and feelings, had already advised him in the early 1890's to get rid of the estate – it would remain a doubtful possession for the sons. But Marie Dohrn resisted such an idea. She saw, not without good cause, a reserve in it which could become useful, once forestry and agriculture were back in good order. The possession gave her a certain personal opportunity and stimulated her eagerness to master a difficult task by her own planning and management.

Besides, she had started a literary work in 1890 which, up to a point, satisfied her striving for independence. Anton Dohrn welcomed it. »Fortunately,« he wrote to Keudell in February of 1891, »she busies herself very seriously and successfully with the translation of two volumes of a historical work of Polish origin – but her own Polish nature needs a larger quantity of external change than our life offers.«

But he calmed his friend some weeks later in a general way: »The Slavic instincts and needs finally submitted to the Germanic nature . . .« The theme would return frequently. The great work she started to translate (which remained unfinished) was by Valerian Kalinka: *The Four Years of Polish Parliament, 1788–91.*[31] The enterprise demanded conscientiousness and patience. Although a Polish mind could be moved by the dramatic end of the State's independence, Kalinska's presentation is a careful, thorough work based on the Archives, whose goal it was to lead the Polish self-awareness away from the state of revolutionary publicity and admiring legend to the sober understanding of national and institutional flaws and personal guilt. The author himself had taken the route from conspirator, by way of literary emigrant, to clergyman of an order, and conservative thinker, and he intended, by setting the truth on a historic basis, to become the educator of his people, who were easily blinded, and blinding themselves. That was also the task which moved the translator, with the result that her fellow countrymen considered her an inconvenient alarmist. One can imagine the devotion of the translator: the work, published by Mittler in Berlin, contains more than 1400 rather small-printed pages; and the translator made the effort to compare, as far as possible, the sources in the archives.

But before the work could be published there were still many difficulties. Marie Dohrn introduced Kalinka's work with a fairly long essay, making known to the reader the personality, life, and points of view of the author, but it is also a testimony to Marie Dohrn's own educated and independent view. With tact and skill the reader is introduced to the motivation through which the Poles view the fate of their nation. That the translator believes in Kalinka's realism is a matter of course. Otherwise, she would not have made the effort to win over the education-hungry Germans to participate in the tragedy of the Polish people and their State.

When the translation was started, a relaxation of the tense relationship between Germany and Poland seemed to be in process, during the era of Caprivi. The new German Chancellor had relinquished Bismarck's politics concerning the Eastern countries. But this reconciliation did not last. The treatment of the Polish question soon acquired a precarious character once again. Keudell said that it was impossible for the book to be published because it endangered the position of the Station in the bureaucracy. The national-liberal leader Hammacher who, by participating in Dohrn's work, had taken over the heritage of old Bennigsen and had changed the relationship to a warm friendship, said the same thing: it was impossible. Kalinka criticized, to nobody's astonishment, the ever-changing politics of Friedrich Wilhelm II, and it was said that the young Empress was exceedingly sensitive to all criticism of the Hohenzollerns. Marie Dohrn foresaw that the work of years had been done in vain and there was a danger of her looking on the situation as a martyrdom. Dohrn would have found that unbearable, so he decided on a step which dumbfounded both his friends and his wife. During a reception at the Court, he mentioned the whole matter to the Emperor himself. He said that, whereas Prussian politics is blamed, Polish politics receives much sharper criticism. Wilhelm II was sensible: a suitable remark in the Preface could say that, and then the work might peacefully be published with the name of Dohrn. The tactics of immediate attack, against which he had been warned, were proven correct again.

Marie Dohrn had finished her work under worsening conditions of health. A bad period started, during which she went from one diagnostician to another and from one cure to another. There were periods of paralysis; the physicians did not know whether they were the results of anemia or whether they were of psychological origin. But at any rate, after 1897 they tied the relatively young woman to a wheel chair for months and years. Sometimes the sickness halted, and there were periods of good health during which longer travels were possible and she suddenly had her elasticity back. But she continued to live under the threat of sudden breakdowns. She had stayed away from the Naples atmosphere since 1902 – her husband also wanted that. But after selling the family's small summer house near Warsaw, Kazimierszofka, she created a home for herself in the Tuscan seaside resort of Forte dei Marmi, where the German-Florentine circle of the Hildebrands and Kurz had settled. Aside from this, she stayed at Wydranka with her old father. Managing this estate occupied her; in never-ceasing work the economic and technical organization was intensified after the German model.

The tensions in the marriage diminished. Where there had earlier been difficulties concerning the education of their sons and decisions on their future, these sons had now grown to independence, and the relationship of open confidence between parents and children now left to the sons the task of finding a well-balanced adjustment. Also, Marie Dohrn began to come close once again to the life work of her husband, to which she had contributed so decisively in the earlier period. She collected and organized documents; thus she learned to see the development and her place in it from a historic point of view. Anton Dohrn was grateful for this. His irritability gave way to periods of understanding gentleness, »all the more«, as he had written earlier to Keudell in a debate on these matters, »since I feel myself to be a rather frail human being, whose occasional successes could only throw a softening light on my sins.«

Thus there occurred beyond the crisis the chance of a new closeness. It was »with real joy« that Dohrn reported to Elise Abbe in the summer of 1907 how a visit in Forte dei Marmi had left him »with a pure, untainted satisfaction. We did not talk either of the past or the future. It was enough that a new present had come about.« And a few weeks later he wrote to the same friend: »Thus

 our life's ways come together again step by step – and I am honestly happy still to be able to experience this end. . .« Over the end of their lives there lies a friendly, merciful reconcilation.

5 ALEXANDER DAVIS

Carl August Dohrn died in the spring of 1892 at the age of 86, almost to the end with unimpaired strength. A few months later his daughter followed him. Anton Dohrn deeply mourned his older sister. To him, as he wrote to his wife, Anna Wendt had been, among his nearest family, »the most intimate human being«, although she »could not cope with her environment«. To his two oldest sons Bux and Wolf, who went to school in Karlsruhe, she had given motherly care. The two deaths were followed by unpleasant settlements with his brothers. To be sure, Carl August had revoked the injustice of his will concerning his oldest and youngest sons and things could have gone smoothly, but the structure of the administration of his estate had been complicated by Heinrich Dohrn. As full of merit as Heinrich was in his communal work, he was just as daring and ultimately unskillful in the way he disposed of the family fortune after the death of his clear-sighted mother. He had invested a large part of the estate in a plantation in Sumatra. That was very romantic and had a rich future for the person who could and was willing to wait – for himself, a bachelor who needed little, it did not matter very much; he had always been ready to put more money into this matter. For Anton Dohrn the whole thing was vexatious. He had organized the Station which had grown with its permanent staff and had secured it for the future, but his private troubles never ended. He had hoped to leave them behind with the help of his inheritance. But that did not materialize. The Soekaranda story (this was the name of the Sumatran enterprise) could not be mobilized for many years. The relationship between the brothers became cool. Heinrich, who unfortunately considered himself a businessman, had to admit his fault. But he did not like to do so.

During this confusing period Anton Dohrn found a new friend, whose active and free friendship became useful for the Station and a blessing for Anton personally. He wrote about this on February 6, 1893 to Keudell, and this account may introduce this man who became so important: »I would like to answer your warm, encouraging letter all the more since you could not guess how Fortuna has once again shown me her favor. As if I were Faust and had expressed the wish to have a magic coat, this is how I have felt for the past ten days. I shall tell you this, with all the details of the new situation.

Two years ago I met a wealthy American, a Major Davis (major from the North American war) whom I considered as very educated and equally high-minded. Of course, the Vaporetto was again the connecting link. We became acquainted and, as it were, friends. He was angry that the United States had no table, rented one for his compatriots and also took on the task of having more tables rented. He is, as I am, 53 years old. He subscribed to the *Fauna and Flora*. For two weeks he visited me daily and showed as much understanding and appreciation for the Station as real friendship to its founder. An occasional, partly charming correspondence kept the relationship alive. Before he left he asked me whether I could advise him about how to use 100000 francs in as productive a manner as the Zoological Station. Innocent as I am, I did not understand the elegant way in which he wanted to donate this sum to me and I promised him to inform him as soon as I had a good idea. He took this to be a delicate rejection of his offer!«

The letter then reports that Davis lay in Cairo, deathly sick, but was now staying in Naples again as a convalescent together with his very pleasant friend, Tilden.

»Do you know what happened in those two weeks? One miracle after another. First, Major Davis decided to show his liking for the Zoological Station by buying a place here. He bought the splendid Villa Floridiana, which Bomba (this is an ironical name for the second to last Bourbon King, Ferdinando II) gave to one of his mistresses as a present. It is situated on top of the Vomero, the most beautiful Villa in Naples with a large park, over 100 years old. I am taking care of the negotiations and have at the moment the power of attorney for everything connected with it. Last year, I tried to make Werner Siemens buy this beautiful place and protect it from destruction. He liked the idea, but he died. Now it is in the hands of a much wealthier man and thus saved for Naples and the world, and I look daily with pride, as an intellectual co-owner, on the marvellous garden. And for what does he intend the park? For every botanical fantasy I want to let loose there. I am supposed to design an orderly plan of how one can create something useful for botanical problems, without destroying the beauty of the park which was created by a German, Denhardt.

That is number one. Now, number two. I talked to him about the danger with which a war would threaten the Zoological Station. He looked through the entire financial situation, bills, etc., and was very pleased with it and firmly promised me that he would take care, in such a case, of the entire Station for two years. Number three. He will build a yacht next year, which will be used for pleasure rides in the Mediterranian, but simultaneously for dredging. It will be larger than the steamer I had planned. It is to be anchored in Naples and will be at my disposal when he does not need it. You know me, Excellency, as a serious man, and will therefore not believe that I am telling a fairy tale. It is all literally true – and the magic coat is here . . . «

There were few enterprises in which Dohrn was involved that kept him as happy for the rest of his life, as the »salvation« of the Floridiana. For its owners had already divided the park – as the growing city came closer – into lots and it was in danger of becoming the victim of speculators. Dohrn's influence, namely that he convinced Davis to buy it, had an effect which surpassed Dohrn's considerations. After Davis' death, the City of Naples itself, with sensible foresight, bought the beautiful grounds and made them into a public park. Dohrn himself did not live to see this, but it would have filled him with great satisfaction. That, in the past, the City administration had given him the park at the seaside to build the Institute, found its counterpart now in the fact that he had saved for the city's future the old park at the Vomero with its irreplaceable trees. This was, as it were, Dohrn's communal master achievement for the city which had begun to be proud of his work.

This late masculine friendship with Davis was never clouded. It caused some trouble: Davis' American companions had left to Dohrn the task of paying attention to the new Neapolitan inhabitant, who treated his not very strong health with insufficient care. His family conditions were also somewhat confused: he lived with his mother and his daughter, and a number of problems requiring Dohrn's tact and diplomatic skills resulted – but being always willing to help, he did not mind them.

Davis admired, with an honest enthusiasm, the German scientist to whom fate had guided him. And Dohrn was grateful to have a man near him who, experienced in business and unselfish, became a good echo for many considerations and worries: ». . . the most noble and discrete modesty combined with complete masculine self-assurance – it is very strange«, he characterized him to Keudell.

The correspondence between the two men shows their growing intimacy. Davis was introduced to the Berlin circle of friends, and his inborn love of music was nurtured and deepened. The American mastered the German language fairly well, and Dohrn attempted to bring the best creations of recent German poetry to his attention. Davis' affinity to the German mind grew so great that for a while he thought of becoming a German citizen. Dohrn made it clear to him that he could be more useful to German matters with his unfettered devotion. With affectionate care the two helped each other in their worries. Davis, too, was subject to exhaustion and at those times it was Dohrn who supported him and appealed to the »Major«, the old soldier in him. And what joy when he could donate some interesting object to the »Floridiana«, which was always open to him and his friends while Davis was away. Ernst Abbe had given Dohrn one of the first telescopes he had constructed, and that now went as an admired oddity to the house on the hill, by which the beauties of the country could be brought close to the house.

When Davis offered his support to Dohrn, it was not meant to be used directly by the Station. He started a fund outside the Station administration at the free disposal of his friend for special tasks, for the co-workers, for the security of health and recreation. From parts of it Dohrn built the handsome house on Ischia – »in majoris Davis gloriam«, as the joke was. He loved this island more than any other place – his point of rest, which he often visited. He loved above all the large, incomparable forest of pine trees, the *Arso,* which had become slightly wild again. The owner, Bourguignon, of Frankfurt, had given it to him, so to speak, for his free use. Dohrn liked to take foreigners there. Jokes were made there: the owner of the Park had set a tablet into a large block of lava which, in a friendly pathetic text, said that this was Anton Dohrn's favorite place – Dohrn's diary notes how during a visit of the Emperor in 1896, Kiderlen-Wächter (with whom, to his own amazement, Dohrn was on good terms) tried to tease him with a pathetic recitation of the inscription. On this island, full of memories, where the Dohrns had spent amicable weeks with Arnold Böcklin almost 25 years ago, the summer place of the Station was built in 1905–06 on the hill San Pietro, above the entrance to the harbor of Ischia. Carlo Sattler of Munich devised the building with an intelligent use of Neapolitan building motifs. On the beach a small laboratory was constructed – if the heat in Naples became too oppressive, work could be done there. The entire construction, which was placed in the beautiful landscape with a sure architectural feeling, served Dohrn and his collaborating friends as a recreational sight; the »Acquario«, as the inhabitants of Ischia called it, became the seat of a relaxed freedom.

But Davis wanted to be active in the field of special scientific problems as well. Although he himself was a dilettante, he was ready to push Dohrn to new problems whenever he himself encountered an unsolved one or received stimulation during conversations with his friend, who was so rich in ideas. Thus, he offered him £ 10 000 in the years after the turn of the century for undertaking certain investigations of the bathymetric distribution of deep sea organisms. As Dohrn recalls: »At that time I had to confess openly to my friend that I could not make immediate use of the generous offer. The personnel of the Zoological Station, starting with its chief and down to the sailors and fishermen, needed a radical rejuvenation before such far-reaching tasks could be undertaken.« These lines occur in a letter by Dohrn to the Chief of the Imperial Cabinet, von Valentini, in February of 1909. In it he also wrote that Davis had promised to put £ 10 000 at the disposal of its founder for the scientific works of the Station.

In his will (he died in 1913), Davis included the sum he had mentioned to Anton Dohrn in the form of a wish (»request«) addressed to his daughter, who was the only heir; she honored it, and a small partial amount, which was paid out, served to organize in 1913 the newly created Oceanographical Department of the Station. Then the war started, and during this time the daughter died. The legacy and the plans disappeared.

6 CONFLICT WITH ALTHOFF

Dohrn's early prophecy materialized – namely, that within a few decades, after his daring beginning, a net of marine stations would cover the world. He surveyed the fulfillment of his forecast with a mixture of satisfaction and worry. Germany and Italy had the least part in the development, their needs were satisfied by the Naples Station, which surpassed all new creations. To be sure, the meeting in 1875 and 1876 of German naturalists and physicians had formed a committee in order to found a state-directed Station on Helgoland. At that time, when his difficulties were greater than his certainties, this bothered Dohrn. But the plan was not carried out. In 1890 Helgoland, which had belonged to England, was taken over by Germany. The old plan was taken up again and found a warm supporter in the Councillor Fr. Althoff. He succeeded in interesting a large German private foundation in the project. Gossler asked Dohrn for his professional opinion; he also debated (as becomes clear in the correspondence with Keudell) whether one could win Dohrn for the position of Director or perhaps his oldest co-worker, with his considerable experience. None of this materialized. Dohrn himself thought that with regard to health and family this was perhaps right, but, since he had worked there for weeks, the overdose of ocean would make him melancholic. His advice was to create a connection with Kiel or Göttingen, but to start on a moderate scale.

The fact that, and the way in which, the Helgoland Station came about, must have appeared to him as a good supplement to his own efforts, particularly since he could not produce the biology of commercial fish, whose research was important specifically for Germany. But he was annoyed that in 1892 the Berlin Aquarium under Otto Hermes started a station in Rovigno at the Istrian coast which belonged to Austria, under the patronage of the Prussian Ministry of Education. Hermes belonged politically to the narrow circle of Eugen Richter, and Dohrn sometimes characterized the state support as a »tip for Richter«. He referred the Ministry of Education to the fact that Frank Balfour had once fought against an English station »in order not to weaken Naples«. Now he saw the strange fact that Prussia was in competition with itself. The suggestion that Rovigno was closer to Germany than Naples he found stupid. The development showed that Rovigno did not gain as great an importance for science as Naples, but took care of a practical specialty which Dohrn had given up, namely to provide inland aquaria with living material. Additionally, instructive courses for students were organized there.

The inner development of the Naples Station seemed to have found a certain moment of rest, after the new building, financed by Italy, was finished, and Dohrn had postponed the laboratory boat. But this was in appearance only. In 1887 Robert Koch had visited the Station and in conversations with him Dohrn conceived the plan of offering his help to the new science of bacteriology. He wrote to the State Department that Naples, as one of the largest ports of the Mediterranean, was a place of infection for cholera, etc. It might also be a home for the new German study of bacteriology. To the Italians, one could explain this as willingness to serve public interests. The Ambassador Count Solms was a little afraid of the matter. He wrote to the State Department that, up to now, Dohrn's inner impulse toward enlargement and perfection had always been successful from a scientific point of view; but the financial situation might be better if he would »start at a somewhat slower pace«. Of course, this consideration had been told to the »maker of projects« earlier. Against this objection he had used (in a letter to Secretary Gossler some years earlier, on November 11, 1883) a very typical formula: »Although I understand that it is my task not to surpass the given financial means, and to avoid the rise of some distrust against my administrative activity, it is nevertheless the result of the *elementary* character of the Zoological Station that it pulls from my hands the reins I try to put on it.«

The »elementary« quality was certainly due to the effort, without any previous example, and the breakthrough to the new sciences, but it was also due to his personality, which, with his plastic imagination (plastic was a favorite word of Dohrn's) saw the perspectives in their fulness and depth. Thus he now turned to the bacteriological laboratory. Southern Italy had gone through some years of serious cholera epidemics, and it could be reported to Berlin that people in Italy understood how much an investigation of the harbor waters would increase the knowledge of hygiene, something that had been completely disregarded up to now. The Italian Department of the Interior and the province of Bari took a table each, although only for two years. With this, one could take care of the first equipment without putting a burden on the rest of the budget. Some works which originated here were already being published, but after four years the attempt which had started in such a lively way was stopped: Italy itself had arranged a new professorship for bacteriology at the University of Naples itself – thus the further financial participation of the country in this new field became uncertain. A special burden on the general budget was inopportune, because it was engaged for some time in other problems.

When, starting in 1879, the Reich gave a subsidy, Keudell had seen to it that Dohrn would receive five percent interest on the capital he had put into the Station, which was figured at 300 000 lire. He had also arranged that a two percent amortization be taken care of, as well as a sum of 5000 lire for travel and representation. In Dohrn's original plans he had not asked for amortization and personal remuneration. But Keudell and the experienced Consul General Beer convinced him of this necessity, since the merely private character of the enterprise had already undergone a certain change. During the first years, Dohrn had renounced the money he received for personal expenses and had used it instead for management expenses. But since 1882 he had used it for its real purpose, to which the officials agreed. The specific task of administration had turned in a different direction since the mid-1880's. During the programmatic announcement in the Reichstag in June 1884, the representative Schlutow, a friend of Dohrn's cause, had seriously pointed out that old age and health insurance had to be created for the life-long scientific officials and the permanent technical personnel. For some years Dohrn had worked on this: »I am beginning to create a pension fund to reward those who stay with us and renounce their academic careers (to du Bois-Reymond, January 6, 1881). The wages, as was frequently both recognized and also criticized during auditing procedures, were amazingly small. The collaborators were expected to be quite idealistic. But idealism would not show up in savings accounts. It was therefore arranged that three percent of the foundation capital must be used to create a pension fund. Besides, a general reserve fund had to be started in case of adversities, and also a fund to renew the steamer.« The yearly contribution of £ 50 which William Siemens had donated for travel costs of English scientists was, with his agreement, used as a health and support fund. The subsidy of the Reich was to run until these funds had reached 300 000, 100 000 and 30 000 lire respectively. The support fund was at that time not limited, but was regulated later.

This beneficial and necessary regulation led in the following years to a somewhat painful dispute. In a Berlin office someone worried that these funds might form part of the basic property and might therefore, within the framework of the present contract, one day fall into the hands of an Italian University. Dohrn fought this idea, but he had to negotiate with the Municipio (which he did not like to do). There, however, he received the interpretation, which calmed down Berlin, that of course these funds did not belong to the property discussed in the contract. This sign of German distrust was so unpleasant for him (what a change!), because now he always found in Naples the greatest readiness to listen to his suggestions and wishes. Thus, he had succeeded, in the summer of 1893, in putting a new paragraph in the contract. As he reported to the State Department, this recognized »my four sons as legitimate heirs and delays the German or Italian University for the entire lifespan of my sons . . . I therefore am now in the position of being able to dispose over the Zoological Station in my will, a success which has great consequences for its future, although it came about very quietly.«

It is not surprising that this far-reaching decision was not immediately understood by all the offices which had to deal with it from either a legal or a factual point of view. The construction of the Institute had such an extraordinary character and was so closely connected with the personal quality and authority of the founder, that a legal scheme concerning the duration of the contract could not simply be taken out of a drawer. Anton Dohrn had started to think about the continuation of the Station and about his successor. Gossler had already suggested that he hand the matter over to one of his sons. This corresponded to his own inmost wish. But who could say anything about the abilities of the young men? The father began to watch his sons and to discuss with them their professional futures and their own inclinations. The second son, Wolf Dohrn, was out of the picture quite early since his talents inclined toward literature. The oldest, Boguslaw, followed his father's interests and chose a combined study of sciences after having practiced for a while in the management of the Station. Nothing could yet be said about the younger sons when Anton Dohrn started to work on the problem.

This had to be done cautiously. Obviously he could encounter unpleasantness and hostility. Dohrn had to make an early effort to settle, within the budget, a certain secure income for the future director, as long as the subsidy would last. The year 1905 had been accepted as the moment in which the funds would be large enough to secure the future. He met with understanding in the State Department. But there was no agreement, either there or in the Department of Finance, and Dohrn himself was not quite clear whether the consequences and possibilities of a »family foundation« would be desirable, which would agree with the Italian conditions of an »Ente morale« (moral entity). He felt that nobody could foresee the scientific problems which would come up in 50 years; but the legal form would complicate the administration and be dangerous for the international character as it had developed under the present form. Those long considerations ended with the decision that the present form should be maintained.

At the beginning of the 1890's Dohrn thought, in the case of his death, of an administrative committee composed of German and Italian scientists. But he did not know whom to suggest for the directorship. In the Department of Finance there was some question of whether the Station would become a financial settlement for the Dohrn family. The person who said this, Councillor Twele, became Dohrn's friendly, reliable supporter as Under Secretary of State, and Dohrn liked to tell and appreciate this. It was not Dohrn's way to accept such an opinion. In a memorandum of January 1898, he proposed the removal of a director's salary of 15 000 lire from the budget. He wrote that he had heard the opinion expressed »that objections might be raised against the Zoological Station's becoming a regular source of income for me and my family«. He continued: »Scientific competence is not enough, it is not even the first consideration. What is needed are qualities which are found much more frequently among men of other professions rather than scientists.« These qualities were: knowledge of languages, experience of the world, skill in negotiations, knowledge of human behavior. Of the many men he thought of – and how many had crossed his path! – he knew only one to whom he would give the directorship without fear, but he was an Italian. And the maintenance of the German quality of a future leader must be a matter of course for all those concerned with the Station. The argument sums up the mistake made concerning the reproach of a family settlement: »On the contrary, I could claim that I am pushing my sons in a professional direction in order to take care of the Zoological Station.«

This dispute, which found a positive solution for Dohrn's own intentions a year and a half later in September 1899, was connected with a long fight which Dohrn had to carry out with Friedrich Althoff.[32] This is an interesting chapter of German scientific politics in which one can observe the war among the various Departments over their domains and their funds. One also sees the contrast between Germany and Prussia, and furthermore the duel of two personalities. On the one hand, the official documents reflect, although with possibly pointed words, the first unimpassioned impression. On the other hand, Dohrn's commentaries extending over years in letters and notes show that he himself understood this phase of the Station's history as a fight for power between himself and the leading man of the Prussian Department of Education.

As long as Gossler, who was at home in many problems of science, led the Department, he was Dohrn's supporter, even against the Chancellor's office. Gossler was dismissed in 1891. Dohrn noted that the reason was an unfriendly remark of Gossler's wife about the young Emperor. A young office helper heard it and passed it on. The next Secretary, von Zedlitz-Trützschler, left no trace in the history of the Station. His successor, Robert Bosse, left the Department of Justice only reluctantly; he had distinguished himself in the service of the Reich in social law-making and did not feel secure in the new tasks. That raised Althoff's position who, under Bosse and his two successors, rose to be the indubitably most important character of the more recent Prussian politics of education.

In 1882 Althoff exchanged the professorship at Strasburg University and his cooperation as the successor of Goeppert in the administration of the young state, for the new task in Berlin as the successor of Goeppert. The first meeting with Dohrn was not unsatisfactory: »Althoff is fantastically enthusiastic about the Station ... in this respect all is well«, Dohrn wrote in a report to Naples on November 14, 1883. But half a year later, in May 1884, he sounded different: »Althoff is very polite ... but he is utterly unreliable and his promises are made of thin air ...« The real problem arose only in the 1890's. It began in September 1892, when the Department of Education asked the State Department whether a further subsidy seemed justified. The latter asked him »to explain these doubts more thoroughly«. The answer came only in June 1893 – the doubts refer to »the positive information which reached me from professional circles about the profitability of the Institute«. The fact is that the Department of Education did not know the details of the contracts, and did not see Dohrn's reports, the monthly examinations by the Consul General, the reports of the Ambassadors or the yearly revision, all of which went only to the Budget Department. What do they have to do with Prussia? But in Prussia they thought: »What does the State Department have to do with biological research?« They asked for the latest budgets in order to take a position in the matter.

The war of the Departments took its time and continued over the years. To what extent Dohrn was informed about its details is hard to say. The Consul General in Naples and the Ambassador in Rome were asked to give authentic judgments about Dohrn's way of life. Mr. von Rekowski made a strong move by describing how sensible and cheap the building of the house was, how simple the hospitality, how modest the traveling expenses. »There is no emphasis on material enjoyments ... Dohrn carefully avoids social obligations which could force him to return them in kind . . .« and so on, up to a characterization of the »simple bourgeois furnishings«.

One can feel the anger of the writer about being bothered with this, and he asked whether the Prussian Department thought that it must look at the rent of the work tables »simply as a subsidy or rather as a benefaction granted to the Prussian scientists, thus standing on a level with the Station's. . . ›do ut des‹« (I give to receive – November, 1896).

This is the salient point. Althoff would have liked to get rid of the four tables. He had the impression that the Station no longer needed the income, and with his budget being exposed to many demands, he wanted to free the money for one of his other purposes, even if it was only a matter of 2000 or 4000 RM. »You know I am stingy«, he said to Dohrn during a visit. But Dohrn looked at it differently: the loss of the Prussian tables at this moment would look like disavowing the Station. He was not worried that the other States of the Reich would follow this action – he could be fairly certain to be supported everywhere by the faculties – but the consideration of how such a reduction would affect other nations concerned him. »What is supported by Berlin,« he wrote to the Secretary of Education on October 20, 1895, »is considered valuable in the world.« It was not only a question of a scientific war, but of national competition. To maintain all Prussian tables was a matter of prestige, particularly with regard to Italy. The Station is certainly in a better financial situation, but it is »no speculation, it puts upon me, as always, personal and material sacrifices, which I can summarize in one word: expatriation.«

Dohrn felt like an advance post of German science in the world. The State Department understood him as such and supported him in every way, but the Department of Education seemed to reject this view. Since he could not make any headway with Althoff, who always assured him of his benevolent interest as was his custom and as was certainly true for the biological problem, he turned directly to Bosse. The conversation was revealing, but gave a new turn to the problem. The secretary was cool. He said, »I am told that you get too much money, you do not need it and you get ten percent interest on your capital.« Dohrn was furious. The conversation took place on February 19, 1896. Dohrn had just visited the Secretary of Finance, Miquel, who had told him that the continuation of the tables, which cost only a small sum, could be assured. Dohrn became angry; this claim did not attack the Station, but himself personally. He would now become the accuser instead of the accused, and he asked for an investigation. He had already said to Althoff that all papers were in the State Department, but Althoff said that he had no access to them and they were not his business anyway. Now a clarification became necessary. He would suggest that the Secretary should hear somebody uninvolved, but a man well-informed on the situation of the Station from the very beginning – Keudell should do this labor of love for him. Bosse could not withdraw. Keudell explained the situation to him. It turned out that the ten percent term had been misunderstood, and nobody had noticed the amortization and the creation of a retirement fund. Later there would be conversations on the question of whether such a retirement fund was a proper expense for science, but for the time being Bosse was restrained. To Dohrn's question of February 20, 1896 he noted in the margin, »I would like to lengthen the contract, and certainly not shorten it, which would lead us only to undesirable enlargements.« The rent of the tables was renewed in its present size but only for one year, and not, as formerly, for three years. Dohrn understood from this that the state of war would continue.

He was not wrong, also in the assumption that Althoff was his enemy. The latter wrote to the Secretary of Finance in November 1897 that, »with regard to other urgent demands« he could not allow a continuation of the present support, »although up to now he had been prevented from lowering the sum by other considerations.« These »other consideration« stemmed from his experience that Dohrn's position in other Departments and at Court was very strong.

Of course, a mind as superior as that of Althoff could not overlook Dohrn's achievements; the latter's creative nature had to appeal to Althoff's similar features. But this strange growth did not fit into Althoff's system of power; and besides, why was it tended with a gardener's love by the other side of the Wilhelmstrasse. If the 8000 Marks for the tables and the thousand Marks for publications were taken away from his budget, they could certainly continue. He only wished not to be hindered in his grand desire to dominate. But Miquel did not wish to create a separate budget for Dohrn. He was of course angry with Althoff, as everybody was. They congratulated Dohrn for not accepting this fact, as well as the entire treatment to which he was subjected, this constant having to wait, these half-acknowledgments; one anecdote followed another. Wasn't it Dohrn's task to fight this battle also for those others who sighed and complained and to show Althoff who was the master, although nobody doubted Althoff's working strength, tactical skill, unselfishness, wealth of ideas, and experience with personnel?

In the diary that Dohrn wrote between the end of 1895 to the end of 1899 and in the letters of that period Althoff was almost the central figure. He almost took over the role Gegenbaur had played in the early and mid-1880's as he valiantly fought for his own world of ideas. Dohrn was not a quarrelsome person, but he felt misunderstood or hurt by Althoff in a fundamental way. His was, despite occasional explosions of temper, a polite and open nature, and he no longer liked to dance attendance, which as a young man he had treated ironically as a technical necessity (and which his father disliked). What was the purpose (when there had been an appointment) of Althoff's keeping him waiting for hours in the waiting room among the many others (and where he met, in the same situation, Gustav Schmoller). The professors might not mind this, but he made a scene. For him, there was no doubt that Althoff was rude and that one had to confront him in the same way.

At that time he was collecting stories which would confirm his opinion; but he also accepted reports by Emil Behring and the Grand Duke of Baden, who made it clear to him that his image of

Althoff was incomplete, that Althoff was not only intelligent and strong, but also far-seeing and helpful. If Althoff were only willing to send his collaborator Schmidt to Naples, whose polite objectivity Dohrn appreciated – he had gone through the Saulus-Paulus fate of official representatives more than once. But that was not going to happen now. Althoff's present young collaborator, Schmidt-Ott, who later became Secretary, would only years later be effective as an active, well-balanced friend of the enterprise. At the present moment he could only mitigate the difficulties and the anger.

When Ambassador von Bülow sent a thorough report in 1897 on the jubilee of the Station, Wilhelm II made a note that this very extensive writing should be put before the Secretary of Education. The Emperor was well informed on the troubles among the Departments. He did not like them – he had a favorable impression of Dohrn's activities, but he was also conscious of the fact that he needed Althoff's creative activity. On his return from a trip to Palestine, in the fall of 1898, he interfered. For a few days Dohrn was his guest on the »Hohenzollern« and the Emperor told him that he wanted to start a meeting of all Departments concerned in order to finally stop the paper war, and then Dohrn would finally know what his situation was. Now Secretary of Education Bosse, who had accompanied the monarch to the East, decided to visit the object of the hostilities himself on the way home. He was treated cautiously, with great hospitality. In Naples, a thorough and well-balanced tradition had been formed on how to teach and simultaneously win over great guests. But obviously Bosse had always had Althoff in mind. »Bosse was here«, Dohrn wrote to a confidant in the State Department, »but he did not tell his impressions to anybody; therefore, there is no clarity about his opinion.«

There could be no quick answer. The meeting the Emperor wanted among the State Department, the Budget Director, and the Secretary of Education took place only on September 8, 1899. Dohrn prepared for it carefully in order to be able to answer objections that might be expected. Every Station founded after his example was equipped with a reasonable salary for the director; when the amortization was over there was no further obligation to pay interest. The income was favorable, rising above all in the Aquarium; but among the work tables there were always some doubtful ones, since new Stations were being founded everywhere. His personal relations and influence were imponderables and certainly not eternal. When, during the meeting, the possible reduction of Prussian tables was once again discussed, Dohrn was supported by both offices. It was clear that this was only a skirmish of retreat. As had been previously discussed with Naples, Dohrn transferred the reserve and pension funds to the administration of the State Department; he withdrew the amortized part of his capital, and received permission to branch off a director's salary from the running income. The continuation of the subsidy of the Reich remained open beyond the year 1905, during which, at the latest, the amortization and the calculated sum of funds would be reached.

There exists no proper protocol on the course of this important meeting. Dohrn was somewhat tired of writing his daily diary at that time. On November 12 he wrote to Abbe »that I won the war brilliantly and Althoff himself offered me peace«. In a few years, he found out that this offer would be loyally kept. But something odd happened: during the first years he was not really happy about the victory and the peace. Something was missing. The idea of having an opponent of great strength – even if he did not evaluate his motives correctly in every detail – nourished his wish to fight and his strategic ability, and enhanced his obstinacy and his will to power. The triumph after seven years was welcome, but it left him somewhat empty.

On a neighboring lot, Dohrn's love of expansion found satisfaction since, during the 1890's, no building or founding had been done, but rather securing and harvesting. The recent German colonialism created possibilities and duties for scientific penetration. A learned, bold and skilful man, Gustav Nachtigal, had begun it. He had visited the Station on several occasions at the time of his research trips which started in North Africa, as Consulate-General in Tunis. Dohrn had a natural sympathy for this kind of person. He followed the colonial development with lively sympathy. In Berlin as well as in Naples, he came to know personally a number of the men who worked in Africa, such as the Denhardts, or in the South Seas, such as Hernsheim, and he liked to listen to their tales. He revered Wissmann and remained in permanent contact with Stuhlmann, who administered the Institute of Amani in East Africa. This type of bold businessman appealed to him as much as the scientists – all the better if commercial and scientific interests met. The meeting with Dr. Carl Peters in February 1896 was more negative; Dohrn noted »an unpleasant impression«.

Whoever visited the Station in 1895 to see remarkable things could find surprises: in the department of the conservator Lobianco he would meet dark-skinned men who were helping with the work. They were Papuas. Richard Parkinson from Holstein, who had for years supervised South Sea plantations and had furthered all research in ethnology and geography, had turned to Dohrn: he wanted to found a Zoological Station in Ralum (New Pommerania) after the Naples pattern. He asked Dohrn to help and advise – he himself could take care of the finances. Dohrn was happy about this, but he realized that if a German scientist were sent there, there would have to be some preparations; otherwise, time, strength, and money would be wasted. Parkinson, who enjoyed the full confidence of the inhabitants, sent him several young people willing to learn. The Colonial section of the State Department collaborated; Dohrn succeeded in finding a scientist suitable for this project, Dahl from Kiel. But the jealousy of the various divisions concerning the directives for work and the distribution of the resulting preparations, etc., later spoiled the pleasure he took in this enterprise, whose beginning he had willingly supported. A few years later he was able to support the East African studies of the biologist Jakob von Uexküll, who had worked in Naples for a long time and had found there the basic features of his theory of the environment.

7 THE JUBILEE[33]

The bureaucratic war with Prussia tortured Dohrn more than a sober consideration of the situation warranted, although in retrospect it also contained some beneficial elements. He no longer needed real »support« as he formerly had. But when he thought of Falk or Gossler, who understood and valued his sacrifices, his devotion, his national achievements, he felt hurt that their successors talked of the good financial course of the Station and looked at it almost as a private enterprise. For a long time the Station possessed, in the international world of science, a rank and dignity of objective greatness, whose necessity and effects could no longer be neglected within the development of the new biology.

This was expected to be expressed in the 25-year Jubilee, partly in order to make matters very clear to some Berlin offices. There was, if this thought should materialize, a choice of certain dates: the year 1870 could be chosen, during which the first contact with the Municipio of Naples occurred. Or 1873, when the first scientists entered the laboratories, or 1874, when Dohrn himself had opened the Aquarium and the whole complex with a fundamental speech. Dohrn, whose thinking was practical, did not want to wait so long. He decided to have the celebration in the Spring of 1897. That meant the end of a quarter century since the beginning of digging the foundation, which had been done without any ceremony. It was not a question of celebrating that day in retrospect, but rather of serving the present day. Friends offered their assistance, and the plan was to add a donation for the faster raising of the pension fund. In the course of the year, however, Dohrn asked to omit this attempt, since some natural catastrophes »had demanded much of those who enjoyed giving.«

As a beginning of the manifold honors which were now to collect on his patient head, Dohrn considered the communication from England in 1895 that Oxford University would give him the Honorary Doctor's degree in the following year. Oxford had entered the table group in Naples only in 1892, almost 20 years after

Cambridge, but hurried now to make up for the delay with quick gratitude. The trip in June 1896 was rather informative. The year had started with the »Krüger-wire« of the Emperor, which Dohrn considered a mistake, since he saw no well-considered political aim in it. Chamberlain, at that time Secretary of the colonies, had more decisively entered the German imagination.

A few months earlier, Dohrn had talked extensively with Empress Friedrich* about the decay of the German-English relations; he was curious to see whether reality confirmed her dark forebodings. He noted about a London party with naturalists on June 22: »After dinner the conversation turned to politics, and I was struck to hear and see how the same negative criticism was voiced about English politics and conditions, as had been the case in Germany about German problems. On the other hand, Wilhelm II is highly esteemed, Bismarck raised to the sky. It would be interesting for Germans to listen to such a conversation. During these discussions I told details of my relationship with Wilhelm II, and it raised the greatest admiration that Wilhelm II even sustained any relation with a man like me. This would be impossible in England which is a great pity. My general impression was that even in the country of the famous political wisdom things do not happen in a way different from ours . . .« And when the conversation, following a discussion of Huxley's position on religion, turned to philosophical questions, he noted that the result was »that the intellectual criticism of all these things starts more from Germany than from England. Nobody expects anything from France.«

This is the mood in which, a few days later, he participated in the ceremonies of promotion at Oxford. Among the nine future Honorary Doctors, apart from Dohrn himself, there were three Americans, but he was mainly interested in two English statesmen with whom he now walked, all in scarlet robes, to the hall of Christ Church: Sir John Morley, the close collaborator and biographer of Gladstone and earlier and later member of liberal cabinets, and

* Widow of Emperor Friedrich III.

Joseph Chamberlain, the new Secretary of the Colonies. One could feel how in this environment sympathies turned more to the founder of the new British Imperialism than to the intelligent representative of classical liberalism. »Chamberlain speaks very well, calmly and surely, with English sarcasm. He is very popular and is applauded wherever he appears.« The mixture of traditional ceremony and spontaneity impressed Dohrn. When he himself went back through the crowd to his place after he was honored, the organist played, to his »great amazement, the Amalfi fishing song, which we had sung so often. I was amazed, and did not understand it.« The Vice Chancellor later made it clear to him: Joseph Joachim, the Berlin violinist to whose admirers Dohrn belonged, had written down the notes of the song and sent them to Oxford asking that someone play the melody during his promotion. The happy idea, which was gladly accepted, pleased Dohrn very much.

Two years later, in connection with a jubilee meeting of the British Association, he also received the honorary doctoral degree from Cambridge. This time he also met many German scientists. He did not pay much attention to the talks; and the dinner speech that he had tried to avoid he had to give anyway. But he enjoyed being able to speak with certain people about his new plans during this international scientific meeting. Perhaps the rapprochement with the French might still materialize? He often spent time with Darwin's sons, renewing memories. Haeckel, too, was among those promoted, and they were both »greeted in lively fashion« during the ceremony – but they did not get together. The days in Cambridge were very strenuous, but they left him with the calm feeling that the Station »is considered a solid foundation all over the world«.

For years, he had been estranged from meetings. During the early years of the Station he had not missed a single meeting of the British Association, in order to promote his cause; he also visited the Meetings of the German Naturalists and Physicians in Innsbruck (1869), in Rostock (1870), and in Breslau (1874).[34] During the latter two meetings he had spoken first about his plan and then quite frankly about the difficulties. But after that he was no longer seen. He had even missed the Berlin meeting of 1886, during which Wilhelm His from Leipzig gave a special talk about Naples and the growing importance of that enterprise. He had learned to think with skepticism of the results of these meetings. The recognition of the Station had become so strong in the scientific world – even if he sometimes felt some personal resistance – that he no longer needed to ask for money from the scientists. All his strength now belonged to financial negotiations, administrative worries, the completion of the whole. But he had, in 1896, let himself be persuaded by his Frankfurt friends to participate in their meeting and had found it worthwhile. There he met Emil Behring, who had discovered the serum therapy for diphtheria in Marburg and had begun a friendly connection which was continued in Naples in a creative way. At that point, Behring made it very clear to him that his opinion of Althoff was erroneous on a human level as well as on a scientific one. Althoff had in fact spoken of Dohrn with great admiration. Dohrn was much impressed by a visit to the Dye Factory in Hoechst. A great number of people greeted him. »I came to the conclusion,« he noted on September 24, »that the visit to such meetings has a special attraction for anyone famous because he gets to see things which are otherwise hard to see. Perhaps this is the reason why some gentlemen visit all meetings with such passion. Of course, vanity is nourished, too, because being privileged decidedly creates well-being. *Principiis obsta* [forget about principles] is the word.«

The »decided well-being« he would be able to enjoy once more a year later, but then it became almost too much. The mass of people who wanted to greet him left him »quite confused«, since »he was after all the only well-known foreigner«. But he accepted the invitation of the American Association for the Advancement of Science, which met in Detroit in August and then he took part, a few days later, in the meeting of the British Association in Toronto – there he also met some of his English acquaintances. The American trip left many impressions, some of them contradictory. In the beginning he considered it a »failure« because the men whose presence he considered important were not there. He felt bored by the poor organization of the meetings – »we do this a hundred times better.« But he was impressed by the speed of the development, of the traffic, and the useful communal arrangements. With a view to Europe he noted, *Caveant Consules* [Consuls, watch out!]. He enjoyed the hospitable country life at Whitman's home in Woods Hole, he liked the charming landscape full of islands, but noted simultaneously that one would one day regret the abuse of the forests. A few times he was asked to talk about the Station – that was expected to be of use for the American biological stations which, although devised in grand style, suffered from a lack of interest and were in danger of being patronized from a political rather than a scientific point of view. This improvisation was successful. People paid him a lot of compliments and found that his appearance would mean great and decisive assistance. He was himself satisfied with his attempt »because I see that my original strength is still there, which after all is part of my nature, so that I can still rely on myself.« Altogether he realized that the trip had been worthwhile: the Station would remain before the eyes of younger colleagues, some of whom he liked very much. And he was somewhat moved that one of these talks on the origin of the vertebrates discussed one of his old theses: »It is amazing how deeply my intellectual work of that time has taken roots. It is certain that it was better than the following ones and therefore shows up from time to time.«

In addition to the two old universities in England, Dohrn was made Honorary Member of two Russian Universities in the Jubilee year of 1897, first in Kasan and then in Moscow, as well as of the medical and scientific faculties of Naples. The Royal Society of Naples and the Swedish Academy also made him a member. The Moscow decision was important for him because it could be helpful in the conflict with Prussia. It also meant the end of the difficulties and jealousies which several times since the foundation of their own Stations had been a threat by some Russian scientists and the

St. Petersburg Department of Education. This need no longer be feared.

Dohrn was no longer insensitive to such acknowledgments. He did not refer them to himself, but rather to a support for the future of the Station; its security had occupied his mind very much since the mid-1890's. This was also the purpose of the Jubilee, whose date was fixed for April 14, 1897. A careful and thorough preparation had introduced it. There should be no lack of honors for ministerial supporters, and it was taking all of Dohrn's skill to prevent these honors from being addressed to himself. He saw to it that his German collaborators received honors from Italy, while some Italians helpful to the Station, as well as Italian scientists and officials, were to be given German honors. Then an address was to be expected – Wilhelm His of Leipzig had taken this on. But in this respect, also, there were considerations which could not be performed without Dohrn's advice and knowledge of the personnel. It had to have an international character. Would the French cooperate, and who would cause them to do so? Finally, everything worked out for the best.

Dohrn collected the speeches in their sequence in a little volume. The senator and anatomist of the University of Rome, F. Todaro, spoke for the Accademia dei Lincei. He had worked at the Station before, in the 1870's, and had often served as a connecting link to the Italian offices, with liveliness and the wish to use his influence. He handed over the address, which contained the signature of almost 2000 scientists and men occupied with the politics of science. Then followed the festival speech by Wilhelm Waldeyer who had been the first scientist to try out the Station's real purpose in the fall of 1873, when it was not yet quite finished. The Mayor of Naples, the Under Secretary of the Navy, and the Italian Secretary of Education came with thanks and good wishes. Then Ambassador von Bülow spoke for the Emperor and the government of the German Reich. Dohrn's answer was a moving look back to the beginning and the gradual development. With generous gratitude he thought above all of his father, from whom he had not only received the first means, but the »intellectual protoplasm« on which the Station had grown. And then he wandered through the daring and the dangers of the enterprise, greeting the men – Germans, Italians, but also the English friends – who had given him advice and support. He was glad that Robert von Keudell was among the guests, and devoted to him, during this festive occasion, a special word of thanks and love. The Italian Navy, to which so much connected him, had anchored a battleship adorned with flags near the Station so that the Neapolitan population would realize that a proud festival was taking place today.

Naples, too, knew how to show its gratitude. The Marchuese di Campolattaro, Mayor of the City, announced, with his best wishes, that the City Council had unanimously decided to make Dohrn an Honorary Citizen. Thoughts – and not only his own – went back 25 years to the day when the edict was issued forbidding the continuation of the building and ordering the tearing down of the finished structure – and now the head of the city was saying that the work »had served the city to its highest and purest honor«. Perhaps this was the most complete satisfaction for Dohrn, that he had achieved total victory on the ground whose difficulties had been described to him by Italians in such a lively fashion. He had conquered a home. For years, the population called him patron of the Aquarium.

In December 1895 Dohrn had begun to write a diary. He had written for years a kind of journal, notes on the receiving and answering of letters with short remarks on their content, altogether a support of his memory. He was often sorry later that, in the restlessness of the early 1870's, he had omitted to take down daily remarks. The diary contains extensive notes on professional conversations, negotiations with ministries, meetings with interesting or uninteresting persons; it contains anecdotes to be told and characterizations to be passed on. It is a colorful mixture: how did he sleep, what work did he do, whom did he talk to, with whom did he play cards or billiards? If he listened to music, he wrote down the sequence and names of the pieces, sometimes adding his opinion on the quality of the performances. He clearly enjoyed this comfortable chatter, but at other times it almost tortured him as a burden which one shakes off with a few unimportant notes. The stimulating new duty he imposed upon himself derived from his reading of the first volume of the notes of King Karl of Romania, recently published. Dohrn had, through some meetings with this monarch, a clear idea of Karl's nature. He felt personally touched by the manifold troubles which the King, who was only a little older than Dohrn, had to overcome with intelligence, courage, and patience when he began his adventures in the Balkan state. The royal notes, which he mentioned to everyone (amazed that so few people knew of them), were probably overrated by Dohrn; from a literary point of view they are completely lacking in charm. But in their almost daily conscientiousness they offer good raw material for discriminating historic investigation. Dohrn may have valued his own writings in a similar way. They were never meant to be published – too much private and unimportant material found its way into them. But aside from being useful in fixing the dates of his organizational and scientific work, the reflections and evaluations of some persons, the political and scientific situations – there exist, if one omits the trifles, on these narrowly written pages many events and opinions characteristic of the period. The writing down, which began on November 30, 1895, and has a few major gaps, stops on December 15, 1899, with the words: »the barometer is rising, the storm is calming down.«

The extent of relationships grew from year to year. The Zoological Station reached almost the status of a small princedom, in which the sovereign received his colleagues – kings, archdukes, archprinces came and went. Now the Serbian Alexander was here, and would leave behind a certain impression of indolence. Talking was easy with the young Ernst Ludwig of Hesse; with Johann Albrecht of Mecklenburg one could talk well about politics. The frequent and long stay of Heinrich of Prussia was a little hard to take; he had good intentions, but a naive way of disposing of Dohrn's time and strength and, except for conversations on the

Navy, not much was said. But that belongs to the job once one has started such a matter. Already as a young man Heinrich had spent weeks there and one had to be nice to him because one needed to cultivate the crown-princely parents. They were also very nice to the old Archduke Carl Alexander of Saxony-Weimar, about whom his contemporaries joked. Dohrn always defended the old man »as usual,« as he once said. He saw in him a remainder of a Goethe tradition, which the old man, although eccentric in manner, kept up gracefully. Dohrn did him favors whenever possible, keeping up a friendly exchange of letters and visits – there were no politics involved and it was quite unselfish. Thüringen had not even a table. In Jena, it was not Carl Alexander but Haeckel who ruled, and he, as the Archduchess related, had permitted a study trip to Messina only under the condition that the student »does not set a foot inside the Zoological Station«. This was no longer important for Dohrn.

Among the men with whom a greater intimacy had now developed was the liberal member of Parliament Freiherr von Stauffenberg. With the death of Emperor Friedrich his future could no longer include major political work. For Dohrn, he remained a part of the world that had disappeared, which was also a fundamental part of his own personality. There was always joy, therefore, when Stauffenberg arrived. Eleonora Duse was a guest, as was Theodore Roosevelt, the American statesman. Then Georg Brandes came from Copenhagen, bringing with him gossip about Ibsen. A serious, somewhat shy man appeared – Dohrn's nephew Georg had sent this stranger. He sang and recited poems; Dohrn was profoundly impressed by him and kept him for days. Starting in 1898, Dohrn became a lively admirer of Ludwig Wüllner, who was accompanied by Georg Dohrn's educated and understanding playing.

The German colony in Naples and its environment grew in those years. The Siemens, too, had rented a house at the Posilippo; the friendship with Werner Siemens grew even firmer through his children, above all Werner's youngest daughter Hertha, who was much attached to Dohrn. He enjoyed influencing the education of the young girl and leading her toward the natural sciences. There is no lack in his diary of ill-humored remarks about the contemporary talk of women's emancipation and the like, but he took it for granted that to an intelligent woman all doors to serious scientific studies should be open. On Capri there was Friedrich Alfred Krupp[35] who tried to come into contact with the Zoological Station. At first, Dohrn did not quite know how to deal with him. »Krupp has snatched up a few crumbs of zoological names« (May, 1898). But Emil Behring, who also settled in Capri, thought that Krupp's interests were serious: »It is worthwhile to take care of him.« That was done. Salvatore Lobianco, above all, had an amicable exchange with the German who was so eager to learn. Krupp hoped to develop his own concept of matters. He was mostly occupied with deep-sea research; he planned to equip a special boat for it. Dohrn reacted in a way similar to the offer of Alexander Davis; would the scientific and organizational forces of the Station be sufficient, since they were strongly absorbed by the current tasks? For the technical problems Krupp was willing to pay – a paper was handed to him, worked out in detail by the Assistants Giesbrecht and Henze according to Dohrn's indications. This contained the plan for a systematic investigation of plankton. Krupp seemed willing to accept this turn of his former plans. His tragic death – shaken by malicious slanders, he died suddenly in 1902 – broke off this carefully prepared plan. Dohrn, who had everywhere given witness to the senselessness of the accusations, was strongly moved by this event. Even though his personal relationship to Krupp had not become intimate, he still had accepted with pleasure his willing help and was glad to have encountered, among the rich Germans, a man of this kind, for whom he had searched in vain three decades earlier. Of course, he now had not only the suggestive faith at his disposal, but his visible success.

Since the 1890's, the summer visits in Germany took on a firm shape. Quite regularly Dohrn spent a few weeks at Keudell's estate in the Neumark, in Hohen-Lübbichow, enjoying the atmosphere of the home and the beautiful forest in which he walked with his old friend. The house in Hökendorf now belonged to Heinrich Dohrn. The relationship between the brothers had become a serious burden when it turned out that the administration of the property had tied down large portions to insecure investments that could not be realized. The anger, however, passed and Anton was glad to find out what Heinrich's honorary work meant for Stettin: aside from the work for the Museum, Heinrich Dohrn deserved great praise for having, as a department head, created the parks by opening up the land next to the Oder River. Anton Dohrn maintained a touching loyalty for his home town (also for the few remaining old teachers of his whom he visited regularly and for the wide forests of Buchheide through which he also wandered). For a while his mind turned over the question of whether he should build a small summer house there, where he would really find complete peace.

The weeks in Berlin which lay between the visits to Keudell and Stettin were full of constant, sometimes frightening restlessness. It seemed unavoidable that he would run from one invitation to another, and also from music to music – he never missed the dress rehearsal in the conservatory when Joachim was playing. He met Joachim in the houses of Mendelssohn and Keudell. He also asked him to let his great art be heard in Naples. The charming anecdote of the Amalfi melody during the Oxford promotion stemmed from such a visit. In this circle he met Johannes Brahms, who was a close friend of Dohrn's brother-in-law Wendt in Karlsruhe. Brahms defended the art of d'Albert whom, as a human being, he did not understand. Dohrn informed the composer of the emperor's inopportune Transvaal wire of 1891. When possible he visited the Empress Friedrich. Their unpleasant topics were always Bismarck and the Emperor, the German-English relationships: »She considers Chamberlain a man of action, for whom she seems to have the greatest sympathy – very characteristic!« (June 13, 1896). In his notes, politics plays a much greater role than science. He dumbfounded members of parliament and diplomats by the well-informed independence of his opinions. Of course he visited his old supporters: Helmholtz was dead, du Bois-Reymond retained his old affection. When he died (at the end of December, 1896), Dohrn lamented »the oldest friend of the Station«. Virchow, in an almost stubborn yet grand sense of duty which moved Dohrn, tried to defy old age. A soft tone came into their relationship which, due to Virchow's sober harshness, had not always been easy. For the old man, who began to be somewhat lonely, the devoted gratitude of the younger man was a source of joy: »the old man was again rather cheerful, he embraced and kissed me« (1898).

Dohrn had a lively exchange in those years with the most influential and best-informed journalist of the capital, Fischer, who represented the *Kölnische Zeitung*. Dohrn owed him much professional advice, since he was well acquainted with the spheres of the various offices. He met Miquel and the new Secretary of the Navy, Tirpitz, who impressed him greatly. The unusual extension of his running back and forth made it possible for Dohrn to meet the national liberal Hammacher and the social democratic leader August Bebel (1896). He could tell him that he had heard him speak once in 1869, together with Abbe, in Jena, and they talked about that man and the socio-political experiment which Abbe had dared to make for some years.

The relationship to this friend of his early, independent years brought him a new, proud as well as caring closeness. At the times of his wild traveling there had been long periods when they did not see each other, although a partly business-like and scientific, partly personal correspondence had always continued. Since the middle of the 1890's, Dohrn had again been visiting him regularly. They exchanged old memories and new plans. To be sure, these visits always left the impression that Abbe was ruining his nerves with his restless work and the enormous amount of business with which he had burdened himself, aside from his brilliant research. Dohrn implored and warned him and tried to lure him to Naples. It is doubtful whether Abbe accepted the invitation of his friend, who was always loaded down with plans himself and often irritable, as the preacher of the kind of life to take care of his own strength. When Dohrn was in Jena he saw that with Abbe, who was just »very cheerful«, there would be no more discussion of a visit in Naples. He tried to characterize the man who was constantly being begged by others to consider sensible relaxation, in a note of June 1899: »He is after all of a naive egotism which he identifies with a sense of duty. He gives up the comforts of life, because basically he does not need them, but he does not give up his habits to please his wife. I had a talk with her about this – she is completely resigned.« Finally, in a sort of pact, Dohrn and Mrs. Abbe succeeded in taking the object of their common worry and love to the South. In the spring of 1903, the Abbes were the guests of Dohrn in Naples for a few weeks and he was happy to show to his oldest friend what really resulted from their conversations in Jena. He enjoyed giving a cheerfully optimistic time to the serious man whose life lay heavy on him. About the true situation, namely that his nerves and heart were used up, he had for years had no illusions: »Abbe pays a heavy price for his almost superhuman wishes and achievements – more than he had ever imagined. As the loyal companion of his powerful creativity you have to pay your part of it. All such pioneers, all precursors, are condemned to a heavy punishment. The highest idealism pays the highest price.«

Thus Dohrn had interpreted, in November 1902, the beginning tragedy of Abbe's hopeless illness. The time came in which one could only wish delivery to the suffering man. When Abbe died in 1905, memories of the past decades broke out in Dohrn. In a letter to Elise Abbe he summarized them: »Both of us put great demands on our strength and reached our goals; how much greater he always was, these goals tell us. Grateful for the loyal friendship of this great friend, I can tell you how much I look up to him and honor him as the greatest personality I have known.«

8 THE GREAT NEW BUILDING[36]

On November 6, 1899, the permanent judicial secretary of the Municipio, Carlo Cammarota, visited the Station. Dohrn also saw this close friend quite often on other occasions. Some time before he had started to look through his old letters, documents from the early periods of the Station, and to make excerpts. As he looked back, he appeared to himself »insane«. As the motto of this writing he chose words of Goethe: »A great intention seems insane in the beginning . . .« During this work, there were also melancholy encounters with some deceased friends – Balfour and Grant were the first, and in the last few years Huxley, Kleinenberg, du Bois-Reymond followed. Among the Italian supporters Panceri and Trinchese were no longer among the living. But in Cammarota he encountered a piece of his own past. This was the man with whom, three decades earlier in March 1870, he had held the first negotiations in the Naples City Hall. When they now met from time to time they talked of the past, since in Dohrn memories had been re-awakened; they spoke of Savarese, Dassi, of the changing gallery of characters who, in that restless period after Italy's union, appeared in public life and disappeared again. Cammarota had remained diligent, clear, and indispensable.

During that fall day, however, the conversation did not belong to the past, but to the future: »One must always have something in front of oneself, then one bypasses oneself«, Dohrn had written to his brother Heinrich some years ago. He always had something »in front of himself« and once even had to swallow the term »miserable creator of projects«. The new plan he presented to his friend of the Municipio he did not consider new because it belonged to the logic of things as written down in his program more than a quarter of a century ago. Cammarota understood this completely. The Station must be enlarged by at least its present size, so that it would satisfy the claims demanded by scientific development. Therefore, a new building lot was needed. Personally, Cammarota agreed to an enlargement toward the east. On November 22, Mayor Summonte visited the building and the suggested lot and agreed, »The dice are thrown.« The unanimous confirmation of the Naples City Council followed on January 24, 1900. Dohrn had already reported with satisfaction to Jena on December 17: »that which once took two years, I achieved now in less than two months without objections.« The financial question remained open: »Under no circumstances will I build before I have collected a sufficient sum – I don't want new calamities.« He let things develop. »I have the place and the plan,« he wrote to the Abbes one year later, »but 300 000 Marks could not be stamped out of the ground, even though a colleague in Berlin recently claimed, in answer to the question of a visiting American, whether I had visited him, ›Dohrn keeps company only with Princes, Excellencies, millionaires.‹ You see the man judged me correctly. I had somebody tell him that he was perfectly right!«

The preparation followed two directions: the financial and the scientific. For it is clear that the work will not be done simply by collecting a building fund. Added to the new laboratories will be larger expenses for personnel and material. The new Ambassador in Rome, Count Wedel, although speaking strong words to the effect that the Station did honor to the German Reich, could not suppress the reflection that Dohrn's plans for a new building would endanger the good financial situation of the present Station. But its leader had believed for some time that he could consider even this with less worry. On February 20, 1898, he had noted: »One hundred ninety American tourists together in the Aquarium.« That little sentence means much. There were at this time several oceanliners which traveled the direct New York to Naples route. Visitors to Italy landed here, and the Station figured among the first European sights next to Vesuvius, Pompeii, and the Blue Grotto; probably also before the Museum. The report to the State Department of 1900 says that »the Americans are replacing the English.« We remember that in the 1870's Dohrn could consider an English Election to Parliament during the traveling season as a threat to his budget! Now such travel agencies as Cook and Stangen and the arrangers of pilgrimages had introduced the Station into their »programs.« More than 1900, Dohrn can report to the office. This is, to be sure, the income from the »Holy Year«. One third of the expenses of the entire Station were covered by the entrance fees to the Aquarium. By emphasizing this, he reminded the office that in the 1870's knowledgeable scientists asked by the German Departments for their expert opinion characterized this construction as a great mistake, responsible for enormous general expense. When Dohrn reported in 1904 that the Aquarium's expenses were 4641 lire, as compared to an income of 77750 lire, he took up the problem once more. Although it could not be said that all costs of the enterprise could be carried by the Aquarium, as he had at the beginning believed, by now it was two-fifths: »Thus science gains from the public a large amount, which could scarcely have been won in any other way.« Added to this was the immaterial effect of teaching and educating. This satisfaction can easily be understood.

What occupied him from the point of view of the politics of science, particularly in his arguments addressed to the Foreign Office, was to preserve the pre-eminence of German science and the Zoological Station. He looked to America and saw the »danger« that there, where »such a brave beginning of problems up to now untouched is common,« i.e., in comparative physiology, a competition with German research was arising. His own studies, which he started again toward the end of the 1890's with a fresh mind, could not follow this turn. One may find that they lack the enthusiastic and self-assured confidence of the earlier approach. He realized that the younger generation, for which he had provided the working place, was turning away almost completely from a genealogical formulation of the problems. Dohrn started again to use the microscope and microtome diligently. He noted with satisfaction that he had succeeded in developing good staining methods for his preparations. A young student, Jordan, who was later to be Professor in Utrecht, was his assistant, and while he trained him, he enjoyed experiencing once again both success and failure. The studies were no longer devoted exclusively to organ formation in the course of the hypothetical change of function, of esophagus, vertebrae, limbs, genital system, etc., but rather to the nervous system, the hypophysis. Not until a few decades later would Science open the door which he had cautiously opened just a crack.

The law of his nature remained, namely that he needed friction, even though it was not his final goal to convince others, but rather to conquer for himself the last certainty. Gegenbaur was now in the background. Dohrn had once told Abbe that »my polemics against Gegenbaur are dictated by the desire to eliminate completely the kind of veneration for his pronouncements which is the most serious hindrance to a healthy development of science. He has reached the point of being allowed to say the most stupid things without anybody daring to say, ›How ridiculous!‹« This goal Dohrn believed he had reached. But the former problematic scientific issues were still not solved. So he turned to a former student of Gegenbaur, Fürbringer. He is, he wrote in November 1900, »of course again on the warpath against former gods.« But it was still not easy to dethrone the old gods. At the same time he kept abreast of new themes in his own environment. He wrote to Davis on July 10, 1903, that he was just reading all publications of Hans Driesch, the leader of the modern vitalists, and that he would like to talk about all these things with his friend: »Both Wilson and I are full of sympathy with the young author whose unusually strict and remarkable powers of reasoning have been working for twelve years to dig a new river bed for morphological thinking and research. I am not taking sides when I say all this, all the more so since he is simply banning my own scientific work to the plane of simple dillettantism – which cannot be taken seriously!! Of course, I do not think of accepting this unsolemn decapitation. One day, I will talk about my own development, but I understand what moves him and I appreciate his point of view and his conclusions. When you return to Naples, I hope I will be able to speak to you about it.« (This letter, which is no longer available, was written originally in English.)

While Dohrn, around the turn of the century, felt again his »old pleasure of fighting« and steeped himself for longer periods into his professional studies, he simultaneously worked on the preparation of his new plan. He had discussed it already in October, 1897, with the most important German chemist, Emil Fischer, in Berlin; he expected much, »although himself a layman,« of the chemical-physiological studies in Naples, since »physiological chemistry had been inconceivably neglected«.

Fischer went along enthusiastically; he was sorry that the efforts up to now were not successful and he promised Dohrn to discuss positively in a letter the problem of need, and to outline the programmatical tasks. This gave Dohrn not only a confirmation of the most important scientist, but also a tool for his future actions. He did not omit to develop his plans before the German physiologists, nor when he visited England in 1898, and whenever else possible. Thus, he created that expectant tension which made the new work appear not as the whim of its founder, but rather as a professional necessity.

The progress of financing had a different character than Dohrn's attempts in the 1870's and even the 1880's. Germany had become richer, the Station was no longer a castle in the air which could be considered as the whim of an enthusiast or the trick of a speculator. It stood firm and world famous on the Gulf of Naples. Dohrn had lived through the old experience that whoever is in need – as he had been at times – asks with less chances than he who is successful; but now all begging had actually been removed from him.

This time it all went without public demand, without meetings, without large committees which had previously missed the goal. Two parliamentarians who were friends, old Hammacher and the liberal socio-politician and industrialist, Richard Roesicke, took it upon themselves to start a quiet collection with good results. When they both, one shortly after the other, died in 1903 and 1904, it was particularly painful for Dohrn that they could not see the result of their loyal services. Andreae of Frankfurt was the legal

counsellor in bank matters. The fourth helper was the Imperial personal physician Leutholdt.

The collection of 1884, which was headed by Gossler, had left 40 000 Marks for the »laboratory boat«, some of the sums had been signed up only conditionally. Andreae had raised the fund to 60 000 Marks through his intelligent administration; and a young English zoologist, Bidder, who believed he owed Dohrn thanks for scientific advice, had added 10 000 Marks. Krupp subscribed, as Dohrn wrote to Davis in the fall of 1901, 100 000 Marks. But he wanted it to be treated quite confidentially, »because otherwise he will be besieged by a hundred other people who all want money«. When the collectors stood before the third hundred thousand, the matter seemed secure enough to draw Wilhelm II into it. The meeting, which took place on October 6, 1901, in the ›Neues Palais‹ in Potsdam, became very important; the Emperor had also asked the Secretary of State von Richthofen and Althoff to join them. On that occasion he wanted to clarify the relationship of the Station to the various offices at the time when the subsidy from the Reich would stop, having reached the amounts needed for the pension and reserve funds – that moment was to come in a few years.

The result was quite satisfactory. The Emperor decided to donate 10 000 Marks from his disposition fund and to permit this decision to be used as publicity during the private requests for donations. That made further efforts easier. The Emperor asked Althoff for his judgment on the plans, »and his opinion, I am glad to emphasize, was favorable and positive«. This Dohrn reported to the State Department in June, 1902. Thus peace was made with the powerful Ministerial Director and the basis was created for the later regulations to which Althoff now agreed as well. The yearly subsidy would continue, particularly since new tasks would come up. Dohrn asked to be allowed to stay within the framework of his connection with the Foreign Office because, as he explained later,

»this was the precondition for the creation of imponderables and of decisive importance for the moral health of the unusual position of the Station in the world. He suggested that the continued subsidy should be reduced, starting in 1905, by one-half, i.e., to 20 000 Marks, and offered the German government ten new work tables in exchange for this continued subsidy. This regulation was agreed to in the negotiations of the following years.

»I will begin to build in October«, he reported to Abbe in June, 1902. »Everything is ready for it. You know that this is no pleasure in Jena, let alone in Naples. But the result will be great progress.« The building was delayed for a longer period than he had anticipated, but the dramatic and comic interludes of 30 years ago were not repeated. He worried, however, about finding the right leaders for the work in progress. The salaries, he wrote to the Foreign Office, were too low for him to find a recognized scientist. Through stipends he would try to get younger men and then »find out which ones are suitable«. In Henze and Burian he found the right men with whom, in the spring of 1906, the new physiological and the chemical laboratories could be opened.

How often and how much rebuilding was done in the whole establishment, mainly in its early stages! The laboratories, the library, the conservation rooms started to expand and wander: steps were added in one place, a bridge over a courtyard in another. A part of this development of the building could still be seen in it later. But the enlargement was erected with great sensitivity for the measurements which Adolf Hildebrand had sketched. Where the new building meets the old one, a central court was left open. The large façade facing the sea, which shimmers through the dark green trees, was crowned along its axis by a strong, reclining figure created by Fritz Behn of Munich.

9 ATTEMPT AT A PORTRAIT

On the medium-sized, slightly obese body, there rests a mighty, bearded head. Its features produce, despite their strong formation,

Anton Dohrn (1899)

a somewhat soft and plastic expression. The first impression might be that of an indolent, yet comfortable, good-natured man. But the eyes are wide awake, asking sharply, forcing obedience from the onlooker, perhaps probing distrustfully. In the full mouth, gently covered by the beard, there is irony and sarcasm. The indolence of the expression is deceptive – there is much hidden tension in this face, which appears during conversations; the man can be a good listener when a story is well told. But he is waiting impatiently for the moment when he can tell a story himself or argue some point. He is not really an orator. The lectures he has to give are carefully prepared, but if he has to improvise he can do that as well – although he was always perplexed that it worked. Not that he is shy; but he cannot quite depend on his temper. How often did it dominate him, in simple discussions or arguments. When that happens in a small circle, it can be made up for. For his vehemence is free from self-righteousness.

When he rises, his heavy, square body contrasts strangely with his almost gracious lightness of movement. There is nothing clumsy about him. A persevering swimmer, a strong hiker – Hans Driesch tells that the pace set by the 60-year-old always left the young people behind. Bowling and billiards are expected to balance the hours at his desk or at the microscope.

When he speaks, his gestures are lively and expressive. Marie Dohrn once noted that she had recommended cigar smoking to learn to control a certain jerkiness of his hands. He became a solid smoker, but without the considered artistry of his father, whose hobby was to select specific brands for certain times of the day.

When he was well on his way, nothing was too much for him. Then his ability to work was extraordinary. So was his strength of quick concentration on something, be it a research problem, a financial decision, or advice about other people's worries. The way he proceeded was somewhat unorganized. Dohrn had excellent ideas about organization, he created regulations, and he also became well-versed in business after having outgrown the naive unconcern of his beginning. In difficult times he calculated household budgets for their private life and was so convinced of the excellence of his tabulations of expenses that he found this educational help not only important for his wife, but also to be recommended to friends. His life was in theory excellently planned, but he did not follow his plan. He complained that people, if they felt like it, stole his time and cheated him of his power to work. He was told that in this case he would have to introduce office hours to protect himself. He realized that, but never did it. If once in a while he very awkwardly threw out someone who came at a bad moment, he noted this almost as a heroic deed, but he would then write an apologetic letter. He was polite and even patient, if he intended to be so; above all he was of an insatiable curiosity about life, if the other person surpassed the merely conventional. His ability to place himself in the other's position was almost unlimited when his feelings were touched; so was his willingness to do so. To a degree, this may have been an unconscious desire to rule, to direct the fates of others even in the realm of their privacy. How much did he carry around with him, giving advice, recommendations, help, whether he took care of people thrown about by life, or whether he wanted and had the chance to be grateful. When he heard that the daughter of the Welshman Lloyd, who years ago had examined the plans for the building of the Aquarium, was in trouble, he took her worries upon himself in a quick and free decision.

His relationship to financial problems went through some vacillations. The consciousness that his grandfather had created a considerable fortune gave to the younger members of the family a certain freedom from care. He took this for granted as a pleasant fact, without abusing it. His personal style of life, particularly clothes, living quarters, food, etc., was always simple; Dohrn could move, with an inborn sureness, among luxury, but did not need it and was without expensive hobbies. In his educational principles there was a certain moral rigor which was not always easy; he did not trust »amusements« which he considered waste, but always wanted composure and direction. He had only a distant relationship to liquor; already as a student he hated the traditional drinking habits. He was no Puritan, but drank with great moderation. But he had to go through many experiences, before he learned how to judge correctly the value and limits of the availability of money, since he always took care of friends and needed so many scientific and literary books.

In its growing years everything was sacrificed to the Station. His caring mother always helped him to overcome crises. But the somewhat Bohemian color of this household, restricted only to private matters, changed in the 1890's: the four sons were in German schools, his wife had medical expenses. Now he started campaigns of thrift as opposed to the previous inevitable debts. Each expense was noted down, and the sons were expected to learn that, too. Not every attack ended with victory.

He could meekly begin a long trip third class and end up in the first class, if a book was so fascinating or his own thoughts preoccupied him so that he had to escape the chatter of the neighbors. It could happen that he went on a trip having given all good admonitions at home: but then there arrived from England or France a shipment of cacti or some other strange and valuable plants Dohrn could not resist. For it had become his late and only passion to collect plants – since the bed bugs of his early youth! – and to adorn the gardens next to his house and in Ischia with unusual plants.

There were discussions about educational logic – that cannot be taken pedantically. He did not want to be considered a philistine or a pedant, and in the Station there was no doubt about his authority. Kleinenberg had been a wrong choice for the task expected of him, and this was not repeated. His view of people had been trained, so was his ability to educate them with a sense of humor and good will, with fair estimation, if necessary also with painful exactitude of individual control. All this had developed into a kind of routine. The task was at times rather difficult. He mastered it in a sovereign manner.

But this was only true of the periods of inner freedom. There were periods – and they never quite ceased – of depression, of irritability, and melancholy. They came over him suddenly, threw him into a feeling of annihilation, independently of the general situation. »For the last three weeks in Berlin,« he wrote to Grant in November 1876, »I had entered again a state of profound melancholy, despite the happy development of my efforts; one of my old, bitter, and painful fantasies again possessed, like a parasite, my entire existence . . .« Fear of death oppressed him. »Fear of death – not of my own existence, no, rather of the consequences of my death for the people I love.« Not only for himself, but also for the collaborators and his family – these were bad times. A dull indolence which he filled with work »as if« it were valuable, but whose lack of creativity he felt subconsciously, furthermore great irritation which became vehement and scared people – all this became habit. He explained to himself that one could grow used to the evil – the doctors had told him that these clouds would come and go and carry no destructive lightning. For a long time he did not trust them, but finally he believed them. His personal letters sometimes reflect this state of both tormenting and being tormented.

When a night of good wholesome sleep is granted him, he notes it with true gratitude. Once, in December 1897, he thoroughly analyzed the feeling of being liberated from this cruel state: »During the morning, gradually a nervous sense of healing penetrated me; and whereas I had, still in bed at 4:30, the blackest ideas which had lasted almost a year with few interruptions and had reached their climax at the end of September, I suddenly had the sensation of the beginning of a real cure and rehabilitation, both physically and psychologically. All thoughts and wishes are working smoothly and easily, serenity and trust have returned and in the evening I talk a lot and have many ideas at the Floridiana. I tell myself again and again that these are not organic disorders, but functional ones – but for months I was unable to realize this myself and I believed with almost objective certainty that I noticed the symptoms of an intellectual decay, softening of the brain, idiotism, and God knows what horrible ideas, which I carried with me for months and almost died of them. And suddenly, Air! And that is called ›malade imaginaire!‹« One has to keep in mind this burden of life when one observes Dohrn, because only then does one find the correct understanding of what he achieved.

In his good times his creativity and receptivity were almost limitless. His *Studies on the Origin of the Vertebrate Body*, which were published in the *Mittheilungen* for over threes decades, would, if put together, fill a series of major volumes. Certainly they were directed toward a definite goal, but by the nature of the subject and by contemporary polemics they are easily diverted, so that they appear unsystematic, although basically they are not.

While remaining specialized in his own research, he always made the effort to keep in view the total field of science. Several times he wrote to Abbe that he, Abbe, knew that Dohrn did not understand anything concerning optical and technical things, but at the same time he asked questions for which Abbe was grateful. He had a lively interest in every new invention. In 1898, Carl von Siemens explained to him the Diesel motor which was just being discussed »which will have a great future«. In 1896 he saw in Paris the first »living« film; he called it the »phonograph of the eye« and looked at the possibilities which are thus given for the visualization of biological things. For a while he planned to write a history of the knowledge of nature and collected some literature for it, made excerpts from ancient writers, concerned himself with the views of Enlightenment and Romanticism – all of this did not go beyond a pile of excerpts. But it remains characteristic that he was looking for this universal balance for his scientific specialization.

The intellectual exchange in conversations and letters was an elementary need for him. Conversation, too, can be a work of art in its free disposal of the mutual positions. To argue, in scientific disputes, in the changing and mobile discussion of theses, in dialectical games, is a passion; one follows it up »until the chalk falls from the walls«, as a young assistant of Dohrn, Gast, once reported. But these intellectual fights were carried out without being disputatious – Dohrn's self-assurance was not narrow, and totally free of over-sensitivity. One of his favorite quotes was by Goethe: »Whoever cannot make fun of himself, is certainly not one of the best.« That was the way he handled himself; he did not reject a good joke at his own expense. He could become angry and vehement, but could also laugh in comfort and ease.

When there was no partner for conversation, he found one by a letter. Dohrn was a passionate letter writer: »I belong to those people who still believe in letters and therefore write the kind of letter which one can read more than once. Business letters are of course necessary – but letters which try to hit the basic attitude of a human being are much more important. This is what I am doing with you, my voluntary ›business‹ with regard to you.« These typical sentences are addressed to the nineteen-year-old son of his friend Keudell. He still »believed in letters« and therefore put his entire personality on those small sheets which he covered with an even, polite, and gracious handwriting. Of course, tone and attitude vary according to the reason of the letter and the recipient. When it is a question of the »necessity« of mere business, he gave up the individual coloring. But otherwise the richness of Dohrn's nature is reflected in them: the gracious joy of conversation, which relates matters like a novelist, the aphoristic point which is formed sententiously, the educational penetration which wants to »hit the basic attitude«. Some of the passages breathe the desire to let the recipient participate in a moment of climax, some enthusiasm, a triumph; but, particularly in his early years, there are also expressions of the fight between revolt and depression.

Diplomatic letters with carefully formulated persuasion exist next to those of complete abandonment or polemic rage – but those were often held back, although many were written. There are monologues, many of which carry the note »not sent off«. In these cases, the writing was an act of freeing himself, of unloading; obviously a necessary inner process to clarify matters or calm himself down. These letters also fulfilled their purpose. In them, Dohrn expresses his greatest strength which he himself in some reflections felt to be his true being: his artistic element. It appeared in his sense of form, ruled in his »plastic imagination« and left its traces in his verbal ability. But not only in language: Theodor Boveri, in his rich and beautiful memorial speech on Dohrn (1910), interpreted the Zoological Station in its planning and construction as a »work of art«. Dohrn himself would have been satisfied with such a view.

He could absolutely not dispense with music. He did not go to concerts – sitting still (which means being narrowed in) was against his nature; but he never missed, if possible, the opportunities of unconventional offerings; at the houses of friends or in his own he always asked for music. He himself gave up piano playing after his youth, because it did not satisfy his very critical demands. But he had an excellent musical memory. When he hiked, he recalled melodies, whistling or humming, and he was glad when he had a partner in this game. This, of course, did not have the genius of reproduction, in which his father excelled until old age – it was only the original expression of an unlimited need for tunes. In a confessional letter to Fr. A. Lange the 26-year-old Dohrn had enumerated Beethoven, Schumann, and Mozart among his »gods«, in whose service he was »very orthodox and intolerant«. To this Olympus he later added Schubert and Brahms. Richard Wagner was not admitted – in this respect Anton remained in the tracks of Carl August, who actually avoided Wagner's music. A note of January 20, 1896, on a Berlin performance of *Tristan and Isolde*, »about which Davis was enthusiastic, I was hypnotized and tired«. The next day he reported a conversation with Keudell, who »at first heard the opera with disinclination, almost bitterness, but then discovered more and more beauty in it«. The judgment of his friend was not unimportant to him. He once went to Bayreuth. There is nothing written on this, but his general attitude of rejection remained untouched.

Compared to his strong need for music, the theater remained in the background. He avoided it almost completely. To visit the theater was for him a very different matter from listening to music. In the four years of his diary, during which many months were

---

308 Anton Dohrn, letter to Alexander Henry Davis, [Naples], April 5, 1898.

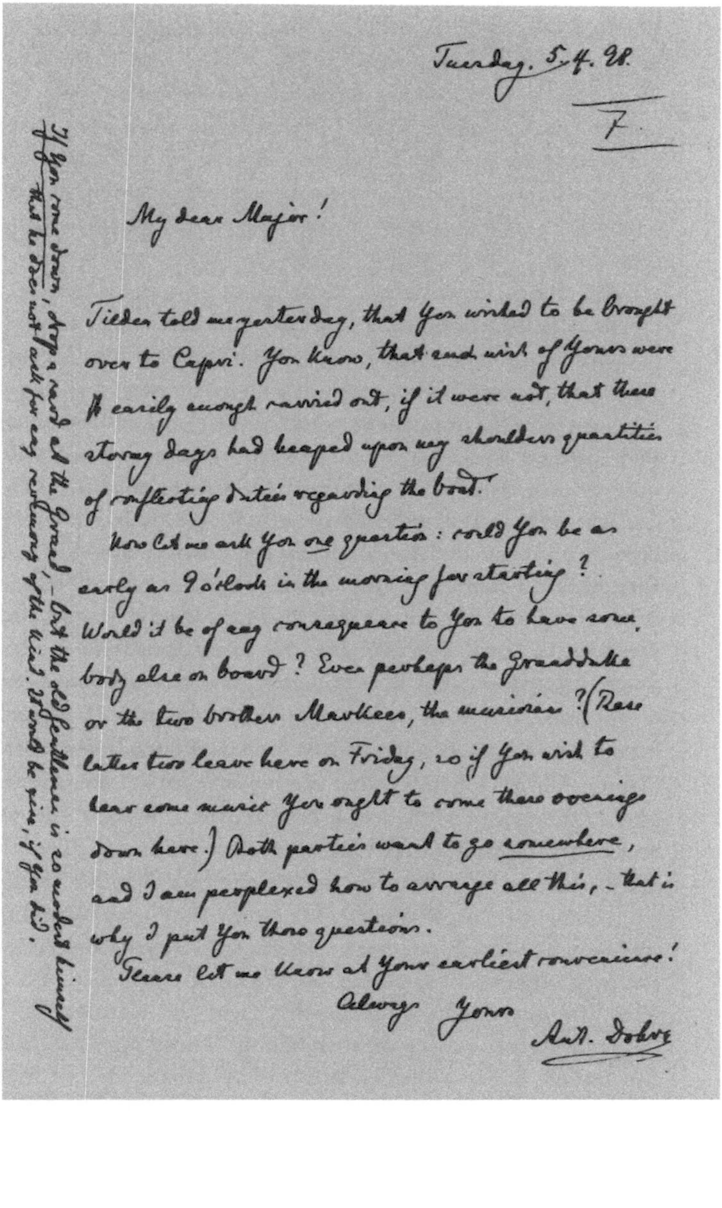

Tuesday. 5.4.98.

F.

My dear Major!

Tielden told me yesterday, that You wished to be brought over to Capri. You know, that each wish of Yours were easily enough carried out, if it were not, that these stormy days had heaped upon my shoulders quantities of conflicting duties regarding the boat.

Now let me ask You one question: could You be as early as 9 o'clock in the morning for starting? Would it be of any consequence to You to have some body else on board? Even perhaps the Grandduke or the two brothers Markees, the musicians? (These latter two leave here on Friday, so if You wish to hear some music You ought to come these evenings down here.) Both parties want to go somewhere, and I am perplexed how to arrange all this, – that is why I put You these questions.

Please let me know at Yours earliest convenience!

Always Yours

Ant. Dohrn

If You come down, drop a card at the Grand, – but the Gentleman is so modest himself that he does not ask for any restriction of the kind. It would be nice, if You did.

spent in Germany, theater is barely mentioned. A performance of *Wallenstein* in Weimar impressed him, but when, in the fall of 1897, he saw Gerhart Hauptmann's *The Sunken Bell* in the »German Theater«, the performance turned in part almost »to pain; particularly the much-praised Kainz I find horrible, since he cannot *speak*, but always declaims in heigthened inspiration and fatal emphasis. If something is supposed to be naturalistic but simultaneously like a fairy tale, such a manner of speaking should be forbidden. That means that I remain removed from modern things.«

The things he remained close to cannot be named in a few words. His letters and notes indicate time and again what reading occupied his mind, some names accompanying him through the years. In his early life, the English novelists played an important role, not only George Eliot, whom Dohrn met several times, but also Fielding and Thackeray. Lawrence Sterne's wonderful *Tristram Shandy* attracted his enthusiasm throughout his life. Don Quixote led the Romance world – there is little talk about Italian literature. Stimulated by Marie Dohrn, Anton read the available Russian and Polish literature: Gogol, Tolstoi, and also Sienkiewicz. Books by Jean Paul, his father's favorite, are mentioned, and also Jeremias Gotthelf and Fontane. He loved Gottfried Keller very much, and often took him up again, and also recommended his works and gave them as presents. The biography of J. Baechtold meant for Dohrn an important key to understanding Keller's humanity. Marie Dohrn felt that the Swiss author knew and moved only in the sphere of the lower bourgeoisie. This, she felt, was his limitation. Dohrn rejected this standard judgment. There is no doubt that, as his sons grew up, he analyzed also the intellectual environment of »new things«. He studied the great Scandinavian writers who were important to the younger generation. He eagerly recommended to Elise Abbe the works of Ellen Key, who for a while occupied him very much. Then he warmed to Selma Lagerlöf's great power of storytelling. These may be accidental encounters, but he collected and read systematically historical political memoirs which formed a key part of his large library. Aside from works on the era of the great revolution and the Prussian revolt, he was mainly wrapped up in the world of Bismarck and his collaborators and enemies since the 1850's. In their autobiographies, Dohrn delved into the great, instructive, intimate, and tense revisions of his own contemporary life. He omitted no diary or correspondence, nothing of importance.

The center of his intellectual life, however, was Goethe. He accompanied Dohrn's development from his early years on, when Dohrn wrote to Huxley in London that he should not miss reading Goethe's correspondence; he himself was collecting everything he could put his hands on. It was the »investigating poet and the poetic investigator« that intrigued Dohrn: »How one could praise Goethe higher than by merely pronouncing his name, in which everything great and greatest is enclosed that can be said by anyone, is beyond my imagination. . . The name evokes the image of the highest perfection of universal humanity.« When he said these words, he did not intend to say anything »new« – it was on May 6, 1903, and at the Palazzo Araniello a plaque was being dedicated in memory of Goethe's stay in Naples. Dohrn gave the speech of dedication, and he did it by confessing his own dedication to the visualizing thinker who was inspired on Italian soil to intuitively define the metamorphosis of the plant and the formation of the skull. Thus Dohrn could visualize his own enterprise as a successor of the great man's, something which »I may boldly say would have excited Goethe's greatest interest«.

To the problems of the fine arts Dohrn had no immediate relationship. His personal friendship with Adolf Hildebrand had taken him to the circle of Marées and Conrad Fiedler; more than once he witnessed, but did not participate in the debates on the elements of the fine arts that were customary in this circle. Although he read and recommended the writings of Fiedler because they seemed to him intelligent and explanatory, his nerve was not touched by them. When Charles Grant stayed with the friends in Florence, in the »old world of painting,« Dohrn admitted that it exerted a »power of satisfaction«, similar to the one he had found, in the summer of

1876, among the lakes and streams, meadows and forests of a »simple nature« in the Polish Kazimierzofka: »I do not believe that I will ever acquire the ability and freedom to find in art what I have in nature – it will not be owned, but acquired. . . If it does not break out powerfully from within, it must be acquired through long habit, if it is to have the power to help people in distress. That I will no longer achieve.« His letters and notes, so rich in reflections, speak much about landscapes, but almost never of the arts of building or sculpture. There is just one typical note of January 1896. In a house in Berlin he is shown a picture by Rembrandt – the self-portrait of 1655 with a large beret and a golden chain: »The painting impresses me profoundly, so that I believe that I have some sensitivity for it; but no galleries, museums, or exhibits.«

When his teachers of the Gymnasium in Stettin had confirmed his lack of knowledge of the Christian doctrine, they added to their otherwise friendly farewell the good wish that »he might find security also in the knowledge of those hings in which the present state of his knowledge had to be characterized as insufficient«. This pious expectation was fulfilled only very incompletely. In Dohrn's youth, i.e., during his dependence on Adolf Stahr and his friendship with Ernst Haeckel, his letters and conversations are full of strong words against sanctimoniousness and the rule of priests. He became somewhat confused, however, by the strange fact that both Darwin and Huxley participated in church life; he noted this with amazement. F. A. Lange freed him of his great materialism. The discipline of Kant and his devotion to Goethe filled his metaphysical need, as much as it had ever existed. The polemic tone against the church disappeared in later years. Life in Italy gave him a strong idea of the positive power of church order and leadership.

Grant's inclination toward Catholicism was certainly without influence, but not without making an impression. And the important position Dohrn held in Naples made him remain in a careful and respectful relationship to the church offices. He saw himself how insufficient his education was for the understanding of ecclesiastical power. He recounted to Minister Bosse an anecdote by Adalbert Falk, who in the early 1870's had led the legislation of the Kulturkampf* heroically, but without much knowledge. The Prussian Secretary of Education had, after he was dismissed, visited Italy and had been in Naples to see the work to which he had given his help and support. There he made the informative remark to Dohrn that »Had I been in Rome earlier, I would have given different laws to Catholicism.« Dohrn took no part in church life. But there was one situation when he had to participate in a church service. It was on board the »Hohenzollern«: »Then we all sang a hymn. How long since I ever heard or sung one! It needed such an occasion.« The Emperor's sermon did little for him; he found its presentation monotonous and without an original idea. He admitted that he »had expected more, perhaps at times something shocking – but it was all rather trivial. If he should send this sermon to the Pope (he had told me earlier that he did this and received from the Pope the encyclical in return) the Pope will smile about his naïveté.«

The letters and notes are very rich in political remarks. It has already been shown, in the description of his life, how he had changed from a radical opponent into a realistic critic of the power relations of the State. He had received congratulations for his engagement from Johann Jacoby, who was in prison and had devoted himself, together with F. A. Lange, to a vehement socialistic reform and a Utopian order for the people. The inner German fights he followed with attention – after all he knew the most important ministerial and parliamentary representatives, but he no longer took them so seriously, and he subordinated them to the great problems of foreign policy. In that respect he had almost

* The term used by Virchow in 1873 for the differences between the Catholic church and the politics of the Reich. It was a fight of many years against the supremacy of the Pope. It ended while Falk was Secretary of Education.

become a professional, although a worrying one, at his post in Naples – through which the entire world wandered, not only scientists. He had close friends among the Italian ministries and knew that their moods and goals were not at all unified. He was in close neighborly contact with Francesco Crispi while the latter lived in Naples; among the later Italian statesmen he most appreciated Giovanni Giolitti's sober professionalism. The many conversations with royal visitors, with their diplomatic and military environment, provided him with a good knowledge, if not of the leading statesmen themselves, at least of the opinions on them which were passed on by various people. His greatly developed talent for treating people correctly, in which his natural self-assurance was as impressive as his great politeness, created for him a great openness during discussions. The image he thus gained of German matters more and more disturbed him in later years, despite the confidence he was willing to have for Bülow's so-called skill. He visualized a dark future for Europe and was tormented by the question of whether and how his life's work could stand up to the wars he saw coming.

From the mid-1890's on he was much occupied with the appearance of Friedrich Naumann. Naumann's religious justification of a political attitude was certainly foreign to him. But in October 1896 Abbe spoke to him in Jena »with great admiration« about this Christian Socialist and gave his writings to Dohrn. Dohrn read them and received a divided impression, that becomes clear in a letter to Elise Abbe: »I have become tired of the minister with his often very pious tone which does not suit him well.« He agreed with the practical demands he encountered and believed that Naumann could achieve more than Bebel, »because he sets himself smaller tasks. I am willing to predict that both directions will thoroughly get into each other's hair and that Bebel would have Naumann hanged if he could.« But Dohrn's »sympathy stops« when he is confronted with the »confused theologian«. He saw the Achilles heel of his propaganda in the theories of Christian socialism.

His attitude changed. Naumann was just ready to free himself from Stoecker's influence, when Dohrn was induced to come to terms with him; now a certain pressure to follow his development was exerted on him by the Abbe household, which untiringly tried to win him over. Also Dohrn's second son, Wolf, who had restlessly experimented in the search for a goal for his life, attached himself entirely to Naumann.[37] His father might think that a decision took place similar to his own a few decades earlier when he met F. A. Lange. »On principle« he declined the Abbes' wish that he should for Naumann's sake step out of his sociopolitical reserve. A letter to Elise of August 1903 is typical of the way he looked on Naumann's flaws: »Wolf goes all out for Naumann who, it appears, finally is in the foreground of public political opinion, where he belongs. Whether he gets into the Landtag or Reichstag as a liberal or a socialist is irrelevant, as long as he finally gets in and gradually learns its practices. I would have liked to see him traveling and living in foreign countries: his quick understanding would have enabled him even more to interfere in foreign policy, as far as public opinion is concerned. What he lacks, as Wolf tells me, is what the French call ›l'habitude du monde‹ [experience of the world]. This may be unimportant, but it becomes important for the person who wants to be and sets out to be influential in all circles, including the top. Bismarck's great versatility and sure manners in the face of the impertinences of foreign »grand seigneurs« rested on this ground, together with his intellectual superiority: if the command of social manners is lacking, often the greatest intellectual strength is not enough to keep one's countenance. Unfortunately, this is one of the greatest German flaws.«

Dohrn was very glad when Naumann entered Parliament in 1907, an event in which, as he reported to Jena, his youngest son Harald participated as »election errand boy«. But he ran into contradictions with Elise Abbe, when the latter approved of Naumann's attack in the question of the Prussian voting rights. He wrote in January 1908: »To be sure, whoever is not interested in the continuation of Bülow's actions might be quite satisfied with the way things go – but I am sorry about the present crisis. I, who never stuck to a political party but rather counted, in all political action, only on tact and success – quite without principle!!! Even Naumann asks, ›Who is Bülow's successor?‹ He should have thought of that earlier!« The three exclamation marks served to scare the receiver, with her definite opinions, in a friendly fashion. At the same time, her emphasis softened the content of the words. Although Dohrn had outgrown all schemas that could have been identified with any party program, he remained free of any trivial opportunism. The passion with which he reacted to social arrogance and confessional intolerance was born of a very firm inner conviction and tradition. But he had learned to distinguish between important and passing matters and laughed or became angry about small quarrels of daily politics which were dressed up as philosophical principles.

10 TOWARD THE END

When the celebration of the Jubilee of the Station in 1897 was followed by a festive evening with several speeches, one of the Italian speakers, Senator Marchese Atenolfi, thought of the future: Boguslav Dohrn should please not act in this realm as is customary among the successors to thrones who start a direction opposite to that of their predecessors: he should rather keep exactly within the directions of his father. That would be best. The 22-year-old student liked hearing this.

Anton Dohrn knew that it could be risky to found some kind of dynasty for a dominion as complex as the Zoological Station. He also was too much of an individualist, i.e., he had too much respect for personal growth to wish to »push«, as he called it, one of his sons into such a task. But he also realized, of course, what an immense gain it would mean for the continuity of the enterprise to have as his successor a man who was familiar, from inside experience, with the atmosphere of the Station, its leading collaborators, and the living conditions of Naples. That might retain the elasticity

Sitting, from left to right: Marie and Anton Dohrn; *standing*, from left to right: Reinhard (Rinaldo), Boguslav (Bux), Wolf and Harald Dohrn. (Around 1906)

of the organization and could counter the danger of a decentralized organizational machinery.

As matters stood now, what was desirable and even necessary was a good overall view of the organization which was just then changing again, rather than a particular scientific passion such as had seized him in his youth under Darwin's influence after his studies had left him cold. But the other part, the »human« one, remained more important.

Boguslav Dohrn had started with the study of medicine and physiology and for a while had done some technical work. His father's letters and notes complain that his son was writing so rarely; he could not quite see what he was doing. But during eventual cooperation in Station matters the young man proved to be skilful. He received special praise for his diving abilities. But the conversations on the future indicate that basically Boguslav wanted to be independent. And in 1899 he decided to say »No« to his father, he wanted to become a farmer. His mother was happy about it – her thoughts and plans were turned to Wydranka, which, until now under her own and the managers' direction, she foresaw under the future management of her oldest son.

The decision of the second son, Wolf, was made outside of science. Anton Dohrn, who looked at the growing young men with proud love, but also with an expectant curiosity, may have discovered features of his own youth most clearly in Wolf's lively and active nature. He objected to the gentle reproach of his Jena friend, Elise Abbe, that he preferred this son. They were all equally close to his heart. But Wolf had the same discussing manner, the willingness to broach philosophic problems. The father suggested to Wolf that it might be wiser to read Friedrich Albert Lange rather than Spinoza, and to Marie Dohrn he explained that she should not worry about Nietzsche's influence on the sons because his philosophy was not directed toward practical ethics. Anton Dohrn saw in Wolf the return of the suggestive influence on people, of which he was conscious in himself, and he was glad of the socio-political elan and quite agreed when Wolf, following the request of Naumann, took over in 1907 the organizing leadership of the newly founded *Deutscher Werkbund* [German Work Society].

Thus the question of succession fell on the third of the brothers, Reinhard Dohrn, who was barely 19 years old when Boguslav said »No«. Harald was still a high-school student; when, a few years later, he decided on a profession, it was for gardening and farming. Ever since his childhood Reinhard had lived with scientific matters, and this is where profession and an inner calling led him. Immediately after this studies he entered the society of the Station in order to take some of the load off his father. Those were the years during which the new building of the Station developed, and in Ischia the summer house and vacation laboratory were started. Also, the laboratories received more visitors than ever before, while his father, sometimes already tired, wanted to complete and resolve with renewed eagerness his scientific problems. His probation went through some difficult situations – it was not easy to develop his own independent profile in the shadow of the superior authority and tradition of his father. But he succeeded with intelligence, tact, and quiet firmness – his father saw it with calm joy. In April, 1907, he wrote to his friend Hubrecht, the Dutch Zoologist, who in the middle 1870's had worked in Naples and had remained a loyal supporter of the Station: »Reinhard is doing very well as administrator and retains his full authority everywhere, because he is a gentleman, not only a professional. That was the problem to be solved for the future!« This was all the more urgent, since Dohrn's health had been subject to more severe attacks since the summer of 1906. Thus he arranged that, from 1907 on, his son be recognized at the offices as his representative with the right of signature. This was acknowledged everywhere without trouble despite Reinhard's youth. In 1909 his father sent him as representative of the Station to the celebration of Darwin's 100th birthday in England, in order to introduce him to the international scientific community. The transition of the direction passed all the more easily, since Dohrn's oldest collaborator and previous representative, Professor Eisig, retired in 1907 after 35 years of service. But he remained in Naples, and his experience and friendly advice were available for the Institute in which he, with his worrying care, had found his life's destiny.

In the past years, Dohrn had come closer to Wilhelm Giesbrecht than to Eisig. In the years which had seen the ups and downs and the back and forth of family worries, and had threatened Dohrn with periods of loneliness, he came closer to his younger colleague. Giesbrecht was a scientifically reliable scholar; his research was devoted to the crustaceans, to which Dohrn had devoted himself in the beginning; he was skilful, full of ideas, agile, and personally modest. As a fine, good observer he was a dependable adviser to all who asked him. But to Dohrn he meant more: the understanding and patient mind to whom he could and did entrust himself completely. He could be sure of his tact, his reticence, his intelligent and conciliatory advice. Among the assistants he was the only married one; his wife was the daughter of the Berlin geologist Justus Roth who, with sarcasm and professional emphasis – whichever the fighting situation demanded – had assisted Dohrn in his efforts to receive help from the Prussian Academy. From that situation evolved the good friendship which tied him from the beginning to Elisabeth Giesbrecht. In his late Naples years she played a major role. She was a woman of sober intelligence and ability, rich more in knowledge than art, lacking real charm, but interested in music – she was willing to pity and at the same time care for the older man she venerated – and he liked this in his times of loneliness. The relationship of the correct North-German woman to Marie Dohrn was and remained difficult – their natures were too different. Dohrn himself overlooked some of her harsh and uncouth features. He welcomed her active care as much as her educated clarity of mind.

Anton Dohrn still wanted to pursue his scientific task, on his »lonely scientific ways, unconcerned about the idea that the entire professorial community considers me insane or something like that«. He had learned for too long to rely entirely on himself to become angry any more than occasionally. He was conscious of his outsider position, sometimes in defiance, sometimes in even-tempered superiority. »Tempo è galantuomo« [time is a gentleman] is an Italian saying – as he had written in July 1902 to his Jena friends, »and if I still have five years of good strength for work and a corresponding calmness of mind, I will bring my great construction of thought under cover and wait until slowly, one after another, they rent a part and then perhaps imagine they have built the whole house, when they only replace a door or wall up a window and open another. I know how much profound work as well as critical and constructive thinking it cost me; and I also know that I have not yet become stupid. Therefore, I consider my mind equally good as that of the others, or – if you don't take this amiss – even better, with all humbleness about the greatness of the work to be done. I never understood *school-like work*, in theoretical as well as practical work I am like a horse trained in freedom: let them wait to see what general result will come of it, when I have to stop in one way or another.« This confidence stayed with him. The *facts* were on his side, he would need only a few years »to present the facts in such a way that the opposition will fall to pieces« (to Hubrecht, March 1905). But this obsessive will to pursue research and proof, which always flared up in the face of resistance, encountered his slow and paralyzing sickness. »At any rate, I have seriously begun to grow old . . . who knows whether this does not also have its attraction?« This was the joking answer to his Dutch friend, who in early summer of 1907 had wanted to take another trip to America with him.

Over the last five years of his life, which had been so rich in work, restlessness, danger, vehemence, disappointment, success, love, and anger, there spread a certain gentleness and wisdom. It is not resignation, the curiosity of life remains alert and the heart of the aging man still experiences with gratitude and a little pain the friendly touch of a late affection. But now he sees new life emerging: the two older sons married: the playing of the grandchildren touched his soul. His nature did not fit very well into the office of a comfortable patriarch, and yet the intimate features of the head of the family and caring counsellor appear more clearly. In his self-imposed »expatriation« he had missed so much of the atmosphere which suited his instincts. His participation is not exclusively directed toward his own sons, but also to the children of his brothers and sister: things had to be evened out or alleviated, and he could enjoy the development of Wilhelm's sons: the musician Georg, and the pharmacologist Max. The strange mixture of the grandfather's inheritance was divided and this division meant the beginning of important achievements.

Among the brothers matters were once again peaceful; the vehement hostilities of the 1890's had paled. When Heinrich was unsuccessful in the election to parliament back in 1893, Anton, as he had written to Keudell, was not particularly unhappy. At that time he was sorrier about the lack of success of his friend who, after his return from Rome, had gone back to parliament and had been a loyal supporter of Caprivi despite his former adherence to Bismarck. But now he observed the eager museum politics of Heinrich for Stettin with sympathy and advisory support. In 1907, with somewhat ironic amazement, Dohrn saw that the almost 70-year-old Heinrich »had suddenly been brought back again by his home town from his political disappearance«, as he wrote to Elise Abbe, and had won back his voting ward for the liberals. »He liked doing it – political life is like alcohol – one can't give it up again.« Heinrich Dohrn found in this renewed duty of the mandate a renewed youthfulness of his somewhat tired strength – he was now one of the few who could tell of the »old times« of the first German Reichstag, and who in their active will represented a fading tradition.

It was a special interest of Dohrn to remain close to his friends' children, to take part in their development, if they opened up to him. Thus he had a strong influence on Hertha Harries, Werner Siemens' youngest daughter, when with a strong hand he guided and helped her with loyal advice on her way to scientific work. He similarly found answers when Keudell's children came to him with questions. Robert von Keudell had died in 1903 at the age of 79. A few years earlier, he had finished his important book on Prince and Princess Bismarck. Dohrn had taken a lively interest in its growth and progress. The yearly visit of several weeks in Hohen-Lübbichow belonged to a firm program, and Dohrn's lively manner made even Keudell, who was very reticent, chat and tell stories. Between the two friends complete frankness reigned. Since Abbe had sunk into his tormenting sickness, and after Keudell's death, Dohrn had nobody to whom he could have opened up in the same way. But it was touching how he cultivated his relationship with the oldest son of his dead friend, Walter von Keudell, first somewhat cautiously in order not to push him, but later with loving openness. The correspondence which the 63-year-old man started with the 19-year-old spreads in part to many-sided long expositions on money matters and estate administration, and is not so important for biography as it is typical of opinions and educational methods.

For instance, when »Uncle Dohrn« was asked about military service, he answered: »I am all for *immediate service* – as far removed from Schwedt as possible. Immatriculation is an illusion and in my opinion a dishonesty, since no one can attend a class during military service . . . but I suggest that you choose a regiment where you have no cousins or other relatives; it makes you more free when you are on your own. Studies in forestry and others will be used much more later than they are now, believe me. One goes into service as half a person and comes out as a whole person – both intellectually and as a man of strong will. And if there is never going to be a war, I would still retain military service as a brilliant education for young men« (August 18, 1904).

It is a piece of intellectual autobiography when Dohrn writes to the young Keudell about his 20th birthday present – the *Memoirs* of King Karl of Romania, which Dohrn esteemed highly: »›Habent sua fata libelli‹ [books have their fate] – it is an old saying which you certainly know, but it can be interpreted in different ways: there are books which have an effect like fate. I know myself that *one* book had a definitive effect on my entire life: Lange's *History of Materialism* – a book you should read after some university years. I read it at the age of twenty-six, and it still influences me and my sons through me. As I judge you now, I believe that no other book could stimulate so much self-reflection as King Karl's autobiography, for he himself wrote the important text of the book.

»It is often said – with a weak glimmer of correctness – that classical education – through the example of ancient men – also forms character. Under certain circumstances and in the hands of individual outstanding teachers this may be the case. But when I first read King Karl's book I immediately felt how much more directly this book penetrates the reader: it is so calm, so objective, and yet throughout so personal, and it is with quite different emphasis that the author of the book and the author of this »book of life« presents himself to anyone who wants to know him.

»I would like to recommend the book particularly to you because I know how eagerly you sometimes have been interested in political matters. You read your father's book, and I believe you also read Dr. Friedjung's book on 1866; and in both you must have felt the difference which such a presentation makes as compared to the daily newspaper chatter or the traditional stump oratory. In the four volumes of the King of Romania you will hear a man who experienced and made a great amount of political history. He is, in a certain sense, as close to you as your father – I mean as a Hohenzollern. In order to feel him closer to you, imagine him as an officer in the dragoon regiment in Berlin, before he went to his great adventure in Romania (I talked to him several times in 1863 during social gatherings at Fanny Lewald's). The entire history of this man up to his coronation will appear to you not as a distant historic event, but rather you will read it with the feeling: ›This could have happened to me – I would have got out of the affair in the same way.‹ Therefore I believe that this book must make a great impression on a young, intellectually lively and strong-willed man – and you are just getting to be 20 years old and beginning to be a man. You may feel the effect of such reading for years, perhaps your whole life. And the effect should not be that you should also become a King somewhere, but rather a man of the iron perseverance, the grandiose sense of duty, the clever subtlety and the far-sighted judgment which everywhere in life conquer or crown when applied in the proper mixture.« Few things show more than this educational letter the noble superiority of the mature man who bent down in loving friendship to the searching, somewhat wilful young man. In its intimacy and simple power it is a moving document.

»With tact and resignation life can well be lived further« even if the forces are only half of what they were, he had written in a letter to

Justus Roth. That was a consoling formula. But it was not always effective. In times of depression – much as he had grown »used« to this evil – anxiety about the danger to life and work overwhelmed him in this troubled hours. But the doctors calmed him. Dohrn wrote extensive notes on a thorough physical examination by Binswanger in Jena, in October 1896: »There is no trace of lesions« (the experiment with the dynamometer had been successful); »in other words, he guarantees with his own scientific name another ten years of undisturbed health. . . Of course, the attacks of neurasthenics or, in my case, cerebrathenics, will reappear from time to time. They were basically connected with inherited states . . . vascular calcification could occur, but up to now there is no trace of it – also this could happen to anybody; otherwise I am to live as always and avoid lasting tensions, emotions, and stress.« This normal advice Dohrn may have accepted skeptically, as any man would who works hard and in eternal emotional strain. Nevertheless he added to this medical information: »Very satisfied and calmed down . . .«

In his promise of ten years of relatively good physical health, against the hypochondriac depression, the great psychiatrist was right, almost too right. At the end of August 1906 Dohrn reported to Hubrecht: »A strange thing has happened. I cannot walk more than a thousand steps. Then I am seized by a pain in the chest, a much accelerated pulse, but insufficient heart beat, and must rest completely.« Was it nerves, heart, or metabolism? »The Aesculapi are not quite clear about the real place of the defect; arteriosclerosis is of course, according to my age, part of it, but they are still searching for another source which would explain the sudden trouble in walking, which is considered as stemming from a neurosis. Since throughout my life I was plagued with neuroses, this is not surprising – only the form is new.« In this collected tone he explained, in February 1907, his tormenting state to his worried Dutch friend.

But does age also »have its attraction«, as he had asked the same friend jokingly? He has no illusions about the situation, with his scientific, unsentimental coolness. One of the doctors who treated him and told him that his patient had become his teacher, admired »in what a masterly fashion you understood how to place yourself in the correct relationship with your defect«. After Binswanger, the great internists Friedrich Müller in Munich and Romberg in Tübingen were consulted for advice and treatment. Cures in Badenweiler and in Kissingen were expected to help – and they did help for short periods. But this existence that had never taken care of itself now had to try to find a different rhythm. He wrote letters, collected documents, and slowly the pages increased on which the stages of the Station's history were written down. This became the essential content of his thoughts; Marie Dohrn was also moved by this duty. She had begun, some years ago in Wydranka, to look through family letters and to write down the passages which treated the worries of herself and her husband concerning the work which they considered their common task.

The Zoological Station caused little trouble in those years. In the line of assistants Gast became an important figure. He also came close to Dohrn on a human level: delicate, sensitive, and grateful, this scientific collaborator also became a nurse during bad days. Some interesting questions came up. Sweden, which had always been so reluctant, now approached the Station; in 1907 the *Institut de France* rented a table for a specialist. Repeatedly, Giard of the Sorbonne had tried again and again to break the ban of the different tradition, and Dohrn's friends in Paris were hoping now for a fundamental change. Dohrn had doubts about this optimism. But the relationship did not completely stop because Prince Roland Bonaparte, who was interested in biology, was willing to take over the table of the Institute. The report of the year 1908 notes that up to now altogether 1858 scientists had worked at the Station, among them 506 Germans. In the last yearly report to the State Department, which Dohrn wrote in 1908, he said: »I am doubly satisfied to indicate here that the laboratories of the Zoological Station never showed so large a number as during this year: including those from the previous year, over 150 scientists worked at the

Zoological Station – a number which surpasses everything I dared to expect.« After these proud words there follows the announcement that his decreasing health forces him »to renounce every continuous activity« – the office should please, in the interests of biology, retain its good will.

In the early summer of 1909 Anton Dohrn traveled to Munich. Marie Dohrn and their youngest son Harald shared the nursing. The heart attacks did not become better, and he suffered much from the changing weather and was anxious, as autumn approached, to return to the south. But this never happened. Quite alert, the patient approached the end. Death came painlessly on September 26, 1909.

The memorial service took place on October 3 in Jena, where Dohrn's scientific work had started. In accordance with his wishes, it was introduced by Beethoven's C-sharp minor quartet, and the quartet in F-minor concluded it. Science took leave of the man who had dedicated a passionate life to its progress, a life which was as triumphant as it was rich in sacrifices: Plate, Driesch, Korschelt, the Danish Bergh, the Hungarian Apáthy, and Hugo Eisig, the oldest friend of Dohrn's fate, spoke for the Station. The ashes were buried at the quiet family cemetery at Hökendorf near Stettin.

A few months later the City and the University of Naples invited the world of science and of public life to the main hall of the University: a great demonstration of gratitude, for which a committee of scientists from all over Italy had been formed, celebrated the memory of the honorary citizen of the city and the creator of the Station. It had been an Italian who had said of the building: »è il Santuario del nostro culto scientifico« [it is the sanctuary of our scientific cult]. In the speech by Monticelli this kind of feeling was expressed.

When the International Congress of Zoologists met in Graz in August of 1910, the meeting was dedicated to the memory of Anton Dohrn.[38] The Würzburg zoologist, Theodor Boveri, had taken over the appraisal of the man and his work – he had won decisive knowledge of his own research at the Naples beach. The close knowledge of the life and activity of the Station gave to his words warmth and understanding. But this was not all: his reflections also contained the measure and rank of the man, the researcher, the strong-willed doer and finally the man who finished it all: »All he had to do was to look around in order to say to himself that he had given to biology an impulse which few can claim, and that his actions and his name would remain a light in the history of our science far out where only the highest mountain-tops are still visible.«

POSTSCRIPT

In 1910 there appeared in the *Publications of the United States' Bureau of Education* a very impressive volume of the American zoologist Charles Atwood Kofoid: *The Biological Stations of Europe*. As an introduction there is a very thorough evaluation of the Naples Institute; the author had received some information by letter. Dohrn did not live to see it, but he would have taken it up with a certain pleasure and looked through it, perhaps not much impressed by the author's very sober way of writing (he simply had to make an official report or »bulletin«). It became something like an address book or an index. But even this much would have awakened in Dohrn many memories: how often he had been asked for advice! And were not most of the Institutes described there children of his mind, stimulated by his Viking-like boldness, his early decision-making and later accomplishment, even if they later turned against him and wished to remain completely independent? His example had generated an extraordinary willingness to follow him, not only because marine biology, which had already been evaluated by Johannes Müller, stepped into the foreground of systematic research, but also because the mutual inspiration demanded an ever-attentive readiness, a methodical refinement, and an extension of the questions. When Dohrn had been driven to his plan by the needs of his own experience, he had before his eyes the difficulties of morphological work, whether it was a question of comparative anatomy or of embryology to which he himself turned first. But as a program he saw the task in a larger context: the public aquarium and the aquariums of the laboratories should make a more thorough research possible of the way of life of the animals. Thus, the turn toward *general* biology received its decisive impulse in the house in Naples. And »nine-tenths of all fundamental work in the field of modern biology« were executed at the Station, as

Hans Driesch said in his obituary, which was divided equally between gratitude and opposition.[39]

Professor Alfred Kühn of the Kaiser-Wilhelm Institute in Berlin-Dahlem offered, in a letter to the author, to provide on request a scientific-historical evaluation of Dohrn: »In the history of biology Dohrn is a typical figure of his time and an unusually effective preparer for the research of the following period of science. He belonged to the few of his generation who took their Darwinism *very* seriously, that is to say they placed their life's work completely in the service of those questions (which were asked but not solved) concerning the principle of change of species and the genealogy of organisms. He was not satisfied to elucidate within a phylum the relationships of the subgroups, genera, and species by comparative studies, or to demonstrate in general or in individual cases that Darwin's principle of natural selection *could* be effective in the transformations of forms. His own research engagement and the foundation of the Zoological Station in Naples were rooted in the conviction that the ›program of the next century of zoological research‹ would be devoted to the physiological problem of the exploration of the actual effect of selection in nature and to the historical problem of the real course of evolution. After he himself had opted for the latter problem, he was attracted by the two questions that were still in the dark: the origin of the two large highly developed phyla, the arthropods and the vertebrates. Of the two ways toward a solution which he as a student of Ernst Haeckel saw as possibilities – the comparison of adult animal forms, or deductions from embryological studies – he chose the latter. He had to choose it, because with respect to the connection of arthropods and vertebrates with lower phyla, which he tried to establish, neither contemporary nor intermediate fossil forms were available. But he was also convinced that »the recapitulation of evolution in the development of individuals« was the more reliable approach. Unfortunately, the two approaches which should complement each other often diverged from each other in individual cases, and for temperamental minds who believed passionately in the possibility of solving the far-reaching evolutionary problems, there arose professional controversies between the comparative anatomists and the embryologists. They became ever more vehement, because none of the adversaries could provide convincing proof from nature for his opinion. In his investigations, Dohrn had established many important facts of the morphology of the crustaceans and the vertebrates. A look into textbooks of comparative embryology detects the traces of his work. Nevertheless, the outsider might find the result of his research meager, because he did not achieve the far-reaching goals, for the sake of which he had started his extensive investigations. There is something tragic about the fact that the problem to which Dohrn dedicated his enormous strength eventually proved to be unsolvable. But it is to Dohrn's merit that he pursued his methodical road with unwavering persistency. The controversy between him and Gegenbaur and his school was very productive because it illuminated the limits of phylogenetic studies altogether. By his uncompromising manner of posing the question, Dohrn contributed to the fulfillment of the general biological task which had been entrusted to the first post-Darwinian generation, and to the creation of the conditions for a new generation which turned away from his problems with equal determination. Boveri, Herbst, Morgan, Harrison, Driesch, von Uexküll, Bethe, von Tschermak and many others came to Naples with entirely new problems. Only through Dohrn's ingenious, far-sighted foundations, which offered not only the wealth of marine animals, but also all resources for the newly developing experimental work in a unique laboratory, could developmental physiology and comparative functional physiology develop so quickly in the last quarter of the past century and the bginning of ours. And only in this way was the leading role in these new fields of research ensured for German scientists. Dohrn's Station, in which outstanding scientists from the entire world were guests, contributed more to the world-wide position of German natural science at that time than any other single institute. More than any congress, it produced personal contacts among researchers and productive connections among directions of research. Even today, under the guidance of Reinhard Dohrn, it cannot be omitted from the life of our science. It is a testimony to the immortality of the problematics to which Dohrn devoted himself and his foundation that for us today, on a different plain, questions about the driving forces of evolution and the history of phylogenetics appear in a new form, out of the past results of developmental physiology and inheritance research. And for attempts at solutions by our generation, the Naples Zoological Station will again play a decisive role.«

This book was devoted to the human being Anton Dohrn and the development of his work, not to the change and the quarrels of the mightily growing science of life. The Zoological Station kept its central rank; one can almost say that, with the enlarged laboratories, it was further enhanced and secured in the years after Dohrn's death. But its fate could not remain untouched by the great historical events.

After his death, the supra-national position of Anton Dohrn was honored on more than one occasion. The memory of the man and the name was intended to remain alive in the gratitude of the science of all nations when a collection brought about the means for a new, larger steamboat; »Johannes Müller« had done its duty for 35 years. The new boat carried the name of the founder of the Station through the world of the bays and islands so close to him, and farther out into the realms of the Mediterranean. In the design, the equipment, and the intention, Reinhard Dohrn went back to his father's thoughts – which had so much occupied his mind around 1880 – concerning the laboratory boat. Now they became, in part, reality: the new goals and tasks were handled with a clear-sighted energy, and special research projects were planned. In April 1913 friends came together in the inner court of the light-colored building where a monument of Anton Dohrn was being dedicated: a fountain had been placed into the wall and on top of it the great bronze relief. Adolf Hildebrand had shaped the mighty thinker's head in stern, self-contained strength. Forty years earlier he had, together with Hans von Marées, enjoyed the happiness of a blessed creation which was certain of its future, free and without worries.

One of Dohrn's worries had always been the fate of the Station during complications of world politics or even war. That was natural: the meaning of the enterprise and the budget of the running and working order relied on a certain solidarity of the will to do research – France's attitude had taught him the political and psychological limits of such an opinion. He had not believed in a conflict between Italy and Germany, although he was no stranger to undercurrents of Italian policy. With some confidence he looked to Giolitti, with uneasiness to Salandra. But to find and fashion a harmony of interests of both states and nations seemed to him a possible and always solvable task. In one of the few cases when he appeared publicly in Naples he described this confidence in a speech dedicated to the devotion of the Italian king.

Five years after Dohrn's death war broke out in Europe; since the early summer of 1915 Italy had stood on the side of Germany's opponents. The work in the Station was not directly hit by the war, but it was paralyzed. Reinhard Dohrn had to leave Naples with his family after having handed over the administration of the Institute to his own and his father's loyal friend, Professor F. Raffaele in Rome. Raffaele had entered the Station in the early 1880's as an assistant, and Dohrn loved him almost like a son and called him jokingly »figlio naturale« [natural son]. Raffaele had taken over the zoological professorship in Palermo since the middle of the 1890's, and in 1914 he was called to Rome. In his chivalrous mind he never forgot how much he owed, on a professional and human level, to his years of learning in Naples; he proved his gratitude in the critical years by his courageous attitude. From Switzerland, Reinhard Dohrn watched and influenced, as far as possible, the development of matters. For a while these took on a threatening character; for the laws of sequestration were applied also to the Station, Dohrn's house, and the place in Ischia; and when the war was over the real problems had just begun. They had a certain innately Italian character: there were persons and groups who believed that the war had eliminated the old contracts which had guaranteed Dohrn's possessions until 1965; but there were many others who, under Raffaele's guidance, passionately fought to take this work out of the formality of a juridical fight. They – scientists, collaborators, or simple fishermen – felt and kept their relationship to the work and its leader with unconditional loyalty. Did not Naples convey honorary citizenship to Anton Dohrn, honoring him and, at the same time itself?

The Department of Education decided, during the time when it was administered by Benedetto Croce, on the return of the Station into Dohrn's hands. But the future regulation was not yet secured. After very troublesome negotiations, an autonomous corporation was created.

The juridical form of the »Ente Morale« [moral entity] was chosen, which means a foundation which consists of an administrative council of five members, presided over by the Mayor of Naples. The Italian Secretary of Education had the power to appoint new members. Reinhard Dohrn became a member of the council and was confirmed as the life-long Director of the Station.

This regulation shows a wise insight into the growth of the Station; it is not pedantic and is generous in practice. The beginning of a new development was given: »The son of the founder has proven himself equal to his father's initiative and his art of understanding situations and handling human beings.« This was written by H. Braus (Würzburg) in 1924 in *Naturwissenschaften*, when he reported to the scientific world that on April 24, 1924, Reinhard Dohrn had again taken over the Directorship and that it was now up to the scientists to recreate and reaffirm the tradition endangered by the world's events.

Thus it happened. To show the history of those years in detail is beyond the goal and meaning of this book. The task was not easy for Reinhard Dohrn. He had to count on new people, and an almost new tradition had to be founded to gain and familiarize assistants and heads of departments. But the father's intellectual and moral inheritance supported his son's strength and obligation;

even more, the past offered the important lesson of the unique laws of growth of the Naples Station.

Anton Dohrn's work, in its nature and quality, cannot be compared with the later foundations which had similar goals. It is not decisive that the house in the Villa Reale was the first modern »research institute«, which, free of teaching and practical tasks, was supposed to serve only to find scientific truth. It is also not decisive that, created by a German mind and supported by Italian understanding, became an international place of stimulation, exchange, supplementation, and noble competition. Our historical view showed time and again that Anton Dohrn, with elasticity and adaptability, behind which there was a large view of a great idea and goal, created an organism in whose essence were the elements of a unique and extraordinary development. For his work it was fortunate that the early well-meaning efforts, for example, of Ambassador von Keudell, to let the State take over the Station to free him of his worries, were rejected by official Berlin. Next to the personal imagination, the worries kept it from any rigidity. If strategy is »a system of accommodation«, then Dohrn was a great strategist. When he started on his adventure, his intellectual expedition, he saw in his mind the final goal and his victory over objections. But the tactics of how to deal with small things as well as the great decisive battles he did not learn from being taught, but from quick decisions or tough perseverance in changing situations. If people considered the Station as a creation of Dohrn's artistic nature, it also represented, if seen independently from his personal conditions, a unique expression of biological laws which his researching mind served: in an elementary way there is nothing made, elaborated, or organized except in the adaptation of functions, in changing conditions, in an almost literal fight for existence. This work formed its character and maintained its leading position while scientific problems and political upheavals changed the world around it.

The preceding passage, written in the summer of 1940,* needs historical confirmation. The second world war brought a relief to the »tactical« situation in which the administration moved: Reinhard Dohrn did not, as in 1915, have to leave the place of his work and responsibility, but was able to remain in Naples. The scientific work continued, although largely restricted to German, Italian, and sometimes also to »neutral« scientists. If there was any danger for the Station during that period, it came from the national-socialist ministerial bureaucracy, which fostered the idea of nationalization, by either persuasion or pressure. To it, it seemed a game of propaganda to include a foreign institute in the scientific politics of the Reich. Also, since the administration of the Kaiser Wilhelm Gesellschaft, not recognizing the special situation of the Station, wanted to harvest where it had not sown, Reinhard Dohrn had to move with great skill within the situation. Fascist Italy had no particular interest in the Station; Mussolini paid no attention to its fate, although his disposition fund sometimes helped out. Basically, men of the quiet opposition, mainly the circle around Benedetto Croce, a friend of Dohrn, had the historical responsibility for the independent continuation of the Station. The Roman ministerial bureaucracy watched – incidentally supported by the German diplomatic representation – to make sure that Mussolini's formula of balance of the »Ente Morale« was not interfered with by the Berlin Department of Education.

The festive celebration of Anton Dohrn's 100th birthday by the university and the city of Naples strengthened the Station's position. A street nearby received the name of its founder. Reinhard Dohrn himself received honorary citizenship in the city, a handsome and dignified recognition of the fact that he was not only heir and administrator of a tradition, but savior and new founder of unlimited possibilities. The position of the family in the enter-

* The book was first published in 1940 at the time when World War II was just beginning.

prise, which was no longer legally private, was professionally and symbolically strengthend by the fact that, by Royal decree, the Dohrn family received a permanent seat and vote in the administrative council and therefore Boguslav Dohrn was called a member for his lifetime. The celebration had a great and solemn character. Mainly the Italian scientists Ghigi of Bologna and Rondoni of Milan praised the work and its historical effect.[40] Also, the home town of Stettin had sent the Director of the natural science museum. A shadow lay over the meeting: Anton's youngest son Harald had not received permission from the political police to travel out of Germany in time. He was already suffering from the denunciations which later became so fateful.

To be sure, the course of the war had not yet given reason for immediate worry. The back and forth of the Libyan campaign did not immediately include Naples, although soon enough the large harbor, from which troops and material were shipped, would begin to attract the attention of the allied air force. That was a warning. Caution demanded that the most precious and irreplaceable possession of the Station, its library of 40 000 items, be put in a safe place. The small town of Pontelandolfo, in the mountainous region of Benevent, would – it seemed certain – never be reached by the war. A textile factory which had been closed became the repository of this treasure in the early summer of 1943. The transportation was a very painful enterprise, but it was high time. The threat soon became more intense: a severe air raid on August 4, 1943, during the noon hour, hit the Mergellina part of the city; the Station itself survived with a certain amount of damage to glass, which had already happened once before from the explosion of a sea mine. But then the terrible news came: the Casa Dohrn on Via Crispi had been hit. The family itself was in Ischia and Sorrent and was unharmed, but the house which Anton Dohrn had erected with the support of Balfour, after the somewhat bold plans by Petersen and through which for more than half a century great European and American guests had wandered, was in shambles – and beneath it had perished three women, trusted companions for many years.

Reinhard Dohrn found a home in Sorrento. In the beginning it was feasible to survey from there the continued development of matters, but the strategic position was badly chosen. For now, after the battle of Salerno in which the allied forces landed in Italy, the great fight for Naples began and connections were cut off. What was happening in Naples? Rumors arrived sometimes, but who could verify them?

Reinhard's oldest daughter, Antonie, who had worked as a doctor in a field hospital until it was moved and had then returned to her parents, wanted to find out. On September 27 she went on foot to Naples and reached the city at a decisive moment. Three days earlier the German commander had announced that the zone of the city along the coast, up to 300 meters in depth, should be vacated and, if necessary, blasted – one of the infamous and senseless actions in which Hitler's war is so rich, and which was intended to impress the simple people or to justify the thin concept of military necessity.

Reinhard Dohrn's representative was his collaborator Giuseppe Montalenti. Whereas between 1915 and 1919 the tough loyalty and diplomacy of F. Raffaele had defended and supported the work, his example was repeated now in an incomparably more difficult situation. The task of Montalenti seemed desperate. It would have been an unheard of chance to find an officer who knew anything about Marées' frescoes, of the »Aquario«, of the history of biological science. What could the urgent adjuration of an Italian scientist mean to him? Montalenti did not even succeed in coming close to one.

Then he met Antonie Dohrn, who had secured the help of an educated German medical officer – and that was the salvation of the Station. For these two succeeded in receiving special treatment for the Station: »For keeping up the Aquarium, a cultural and research institute founded by the Germans«, as it is said in the special permission of September 27, 1943. Montalenti and the technician Della Morte could now enter the closed-off zone. For the first time in its history, the Station had been without custody and care – but the pumps still functioned, and that was the main thing. Of course, there were turbulent days still to come: the revolt of the population against the slowly retiring German troops still brought some danger as the fighting approached the Vomero. On September 30 the skirmishes of the retreating Germans moved away from the city limits, but the work of Anton Dohrn in its exterior and the most beautiful monument of great German fresco-painting were saved from destruction.

But this was not the end of the worries. On October 1 the allied troops moved in. Montalenti was not a little scared when he found a poster on the Station: »Reserved for the fifth Army.« While he wondered what to do next, a few words written with pencil on the wall caught his eye: »In loyalty a last greeting« and the name of the helpful German medical officer. Would he have the luck to find among the foreign soldiers someone whose attitude was similar to that of this German? He was lucky. Although the Station was claimed as a medical post, as living quarters for some officers, it was spared after the change of troops: on May 1, 1944 it was completely returned to its original purpose.

But what had happened to the library? The connection with Pontelandolfo had been interrupted for a long time. The radio broadcast of October 1943 announced that the town had been reached and occupied by the allied advance troops. One of the employees of the Station had been placed there to watch it, and what joy there was when he now appeared, having walked the 80 kilometers on foot. The library, he said, had been untouched. But would it remain so?

In that town a recreational center for English and later also for Indian troops had been established – soldiers very rarely feel the desire to exercise the office of conservators in the midst of their destructive mission. The allied military government was favorable to the request: it helped in the return transport of the books. Only some twenty of the 40 000 volumes were missing – a miracle. Also, the losses among the precision apparatus and optical instruments remained within very minor limits.

It took some time, of course, until the good news reached the friends in Germany. But a message by way of Lisbon arrived, although specially concealed in coded phrases according to the necessities of that time. The old English zoologist, G. P. Bidder, who had in 1900 donated a fairly large sum for the enlargement of the Station, out of gratitude to Dohrn, now wrote a letter to the *Times*, speaking of Anton Dohrn's achievement for the world and recommending his work to the special care of the allied military command. That was a moving indication that gratitude and a fair mind had not died. To speak of this action with praise in Germany would at that time have been considered treason.

Bidder's words were not in vain. It must have meant great happiness to Reinhard Dohrn when he received the news, in the winter of 1943–44, that the venerable Royal Society of London had granted to the Station the extraordinary help of a £ 1000! That was an immediate support for the financial situation of the Station, which was still completely in the dark. How much greater was its moral support – the war was in its hardest phase. At the end of 1944, Switzerland and Sweden (early in 1945) renewed their work tables, and after the end of the fighting the same news arrived from Oxford and Cambridge and the American National Research Council. Also, the Italian offices were helpful. At first they were dependent on the Municipio, but after the restitution of contact with Rome in June 1944 they could again take up the connection with the Department of Education and the Consiglio Nazionale delle Ricerche. In the Station a »Centro di Studio per la Biologia« was founded under the directorship of G. Reverberi. Thus, the messages which Reinhard Dohrn sent to Germany, although reporting also of sicknesses and trouble, were happier than could have been expected earlier.

Reinhard's son Peter, after studying in Germany, England and Italy and participating, as a doctor, in the final phases of the war in the Alps, had luckily come through all dangers safely and now came to the help of his father. But what they heard from home was depressing. The major development went, as one had foreseen for years without illusion; only one question tormented the patriotic feelings even in the South: whether the sacrifice of German history – and perhaps of the German future – would be carried out up to the last insanity.

They heard through the war reports that the Russian lines had moved from the East and Southeast to Stettin and Hökendorf – where the graves of father and forefathers lay in the old, beautiful park – would that place, so rich in memories, be swallowed up by the war and probably annihilated? Where could thoughts of the oldest brother Boguslav turn? Was he already in Thuringia, where he had reported a retreat? He had been there, but was now on the move once again, evading the war and traveling to Bavaria – for many long months there was no news.

And then came terrible news. Harald Dohrn had been arrested early in the summer of 1943; his son-in-law Christoph Probst had, after Stalinigrad, participated in the Munich student revolt together with the Scholl family in Ulm and was executed. The criminal proceedings against Harald were closed after three months of detention, but when the war approached upper Bavaria in April 1945 the SS took him over in Wiessee and brought him to the Munich Gauleitung, together with his brother-in-law, Ministerial Councillor Hans Quecke – who had accidentally been there with him. They were transported further, and finally both men were shot without trial together with others in the Parla Forest. Months later, in the middle of June, his corpse was found.

Hardly any family of those who resisted National Socialism in those evil years was spared tragedies such as this. Harald Dohrn had given up the humanitarian tradition of the family in favor of a lively Catholic faith which had given security and balance to his searching, adventurous restlessness – a melancholy end for a son of this family which was so rich in temperaments and talents.

The first edition of this biography ended thus: »The fact that Anton Dohrn's work emerged from all this unharmed is an experience which permits an optimistic view into the future at a time when a historical look backward must be closed during another disturbed period.« This sentence seemed to us, in the dark hours of those years, to be somewhat too optimistic in its anticipation of the future fulfillment of wishes and expectations. But let it stand as a testimony to all men of good will who can be moved by this example of an achievement which served the world.

THE HOUSE ON RIONE AMEDO

REMINISCENCES OF MARGRET BOVERI

August, 1943

On the street, the afternoon heat of the first hot spring days, the squeaking of the streetcar at the switch, the sleepy, monotonous call of a street vendor; in the house, the opening of the heavy door, the coolness of the entry hall, the calming dark colors of the little library, leather binding after leather binding on eight narrow walls up to the ceiling; stepping out on the narrow iron girders: the glittering light of the gulf of Naples, tender green leaves and at my feet a sweet-smelling cover of light purple flowers; hovering in the air, many hundreds of scents mingling with each other in a dense undulating web of tendrils with which the old wisteria spread a roof over the lower garden.

In this sequence of impressions, innumerable people came to know the house at the Rione Amedeo for the first time. There are few who could resist the magic of this house, its serenity and melancholy which would last for decades. Whoever was a guest here only once felt that he was in a house in which past and present, southern ligthness of living and northern initiative, eastern intensity of feeling and German thoughtfulness ruled equally over the minds, even if he did not know that the picture of the bearded man in the study had been painted by Marées, that at the piano in the music room the most famous musicians of Germany had sat ever since the early days of the house, and even though he could not see what documents and letters were hidden in the brown cupboards.

The guests, however, who were invited more than once to the wisteria-tea soon discovered that almost everything in this house was different from normal houses – whether German or Italian

– and that its structure and way of life could not be measured by other comparable ones but only in relation to itself. The amazement would start when the bell rang for the first time for a meal. For, although one had walked on even ground from the Rione Amedeo to the library, suddenly the family walked downstairs on winding marble stairs. Downstairs a door opened – not into the basement, but once again onto solid ground, into an airy small dining room with a long table. In front of it was a walled-in garden with palms, camellias, mimosas, and yucca, over which during the summer months twigs and leaves of the wisteria would lie like a dense roof so that even on the hottest days a mild, green-golden twilight existed. The house was standing on a steep incline, and therefore garden, dining room, playroom, and kitchen were one floor below.

Rione Amedeo was the name of the street at the end of the 1870's, when the city advanced with only a few houses on this slope in the west. All around the area was clear, and a little further down was the convent, isolated in its large garden; below and next to it there were narrow old lanes, steep or straight or going downward at an angle, in which at night the smell of brown coal fires, little corn-cakes baked in oil and fish, hovered. Farther down there was the wide, splendid Via Caracciolo burning in the sun, which formed the stone wall against the sea. The waves of the Gulf of Naples hit the open beach between Santa Lucia and the little fishing port of Mergellina and the fishing boats pulled high up. The conservative Neapolitans still called the street Rione Amedeo after it had been called Via Crispi for years, when on both sides villas and rental property had been built and the last open spaces in front of the convent had been filled up. Even the inhabitants and the friends of the family said »Rione Amedeo« when they got into a carriage or a taxi. If they had used the official name, they would have been immediately recognized as strangers and everyone living there would have found it appropriate to increase the price a little.

The light-colored, high house with its view of the sea and the distant contour of Capri, the Sorrento peninsula, and Mount Vesuvius had not been invented by an architect but by a young and daring scientist who wanted to give a home to his family after he had succeeded – despite all doubts of the professionals and all obstacles of uncomprehending officials – in creating for his science a work place by the sea. Anton Dohrn was the name of the bearded private professor from Pomerania, with his broad forehead under the dark broad-brimmed felt hat, whom Marées had painted in the group of his friends on one of the frescoes of his recently finished Zoological Station. During the erection of this institute and the Aquarium belonging to it in the most beautiful public park of Naples, the building officials of the city had had more than one word to say and really had done so, to the despair of the owner. During the building of the house at the Rione Amedeo the building police seemed to have shown little interest; otherwise they would have found out, during the building process, that one staircase was missing, i.e., the inner staircase just described, from the »even ground« at the level of the Rione Amedeo to the »even ground« of the garden. It was added later. This created, in the planned classic symmetry, some angles and flaws which were immediately used for special purposes and played a lasting role in the life of the house and its inhabitants. There were several such odd things, partly planned from the beginning (such as the movable wooden wall on a bedroom floor), partly happening through the exhausting climate of the Gulf and the nature of the inhabitants. Not only the owner, but also his unusual wife and the growing sons had created their particularities. The third son took over, in difficult times, the leadership of the Zoological Station and the house.

Aside from three generations of the family, some odd characters, both male and female, had been living in the house for a longer or shorter time. There was Hermann Linden, who as a young man had read a report about the foundation of the Zoological Station and had thereupon written to Dohrn that he wanted to take over the correspondence, book keeping, statistics, etc., the entire bureaucracy of the unusual scientific institute, which was self-supporting. Linden's favorite place was the volcanic island Stromboli;

his passion was the construction of a boat which was supposed to be moved not by the strength of the wind, but by that of the waves; his social pleasure was to mumble to himself interesting stories in a barely audible voice so that the fascinated audience could follow him only with the greatest strain. He had introduced iron order into the archives, finances, registers, and correspondence of the Institute as well as of the house in the Rione Amedeo, a system still alive in its original rigor in the hands of his fourth female successor, as if the little man stood behind her growling some half understandable words about the negligence of the later generations.

Particularities such as those of Hermann Linden were, with the vital persons in the house on Rione Amedeo, not felt as a source of irritation, but cultivated with a kind of amused love. It belonged to the peculiarities of the house that, despite the passions of feeling at all times alive in it, there was a tolerance toward inner and outer problems, a tolerance whose roots were not in an effort to equalize things but rather to enjoy the richness of human beings. But the strength to reject unrelated persons was never lost.

Not only scientists were guests in the house, coming from four continents in order to make their experiments under unique material and intellectual conditions. To the mind that looks back, it seems as if of all travelers who wanted to »see Naples and die« always the most open-minded and creative came to the house on the Rione Amedeo. High State officials, industrialists, and bankers, who had been helpful since the foundation and building of the Zoological Station; princes from many countries who had become patrons, archeologists who were returning from their work in Egypt, Mesopotamia, or Constantinople; diplomats who were going by boat to their new posts at the Mediterranian; writers, actors, painters who were looking to the south for recreation or work. One morning a member of a Royal house sat in the early morning in front of the oval table in the little library and had the lady of the house explain to him how one might kill the numerous fleas of the southern land – for even Royal Highnesses were exposed to this plague when fleas had not yet been exterminated. Werner Siemens,

Eleonora Duse, the Ambassadors von Keudell and von Bülow, Fridtjof Nansen and Ludwig Wüllner, Ernst Abbe and Rudolf G. Binding – to name only a few of the deceased – were among the people who kept their friendship with the family.

But it was not a procession of »prominent people« who were »collected« there. Mutual sympathy was always the decisive matter in the choice of friends. Many a young man without a name was taken in and received decisive impulses for his entire life – in conversations, in looking at books and pictures, above all in the amazing atmosphere of the house. Whoever was able brought something or left something behind to express part of his gratitude – perhaps a painting of the island of Ischia, or the latest book which had moved him, or, after the destructions and losses of World War I, the furniture of the room on the highest, airy terrace floor. Thus the house always gave the impression of being a center into which streamed, through its own strength, the best and latest, but also the oldest objects from various countries. And yet no artwork was professionally collected, no one bought Renoirs, or Cézannes, or china. And if there was a bust by Adolf Hildebrand in the music room, the reason was that the sculptor had one day been seized by the desire to create the head of the second mistress of the house.

On what the effect of the »Casa Dohrn« really rested was never captured in words. It was not only the human beings which caused this effect; it was not only the house that seemed to exude so much of what three generations had put into it. More than once the friends tried to pin down in conversation where the secret lay. It has also been tried in festive speeches, articles, and books. It was never quite successful. No summary of objects or qualities, not the memory of the atmosphere during an evening concert, no picture of the grace and the high spirits which came about when costumes were brought out and there was dancing – none of this can give an idea of what, for want of a better term, was always called »atmosphere«. For it was always the humanity behind the music, the dance, the serious conversation or the suddenly appearing smile of the last owner, which clinched the matter.

We are writing in the past tense. For the house and the wisteria, the library chairs which, under the changes of humidity and dryness of climate often became unjointed, the pictures and rugs and books and pottery are now ruins and ashes. We do not know how many houses still stand on the very peaceful Rione Amedeo. »House destroyed« were the first two words of the wire which brought the evil news a few days after the last destructive air raid on Naples. Certainly, the memory of the house will remain forever in the hearts of those who loved it. Certainly many Englishmen and some Americans, who had once taken part in the wisteria afternoons, may have sadly wondered about the Rione Amedeo when they read that »their« bombs had attacked Naples »successfully«. Certainly the friends in more than one European country may have called each other, without being able to give any consolation, because they did not know how they might, in this hour of loss, express to the family their love and willingness to help. But the cold words remain: »house destroyed«. And only the conviction that what lived in the house, more than stone and mortar, more than the loveliest, collected traditions, more than the effect of the individual members of the family, who since the founding days of grandfather Anton seemed to live an advance life of »Europe« – only this conviction permits the attempt to dedicate an obituary to the house, something normally given only to human beings.

LETTERS BY ANTON DOHRN

This description of Anton Dohrn's life has often drawn on the letters with which he accompanied his work. Some letters were also printed in their entirety in the text of his biography, such as those, for example, that related the story or the characterization of his American friend Major Davis (p. 262), or reflected o the autobiography of King Karl of Romania (p. 323). The added appendix has been kept short due to the lack of space. Only a few items were chosen, which may serve to illustrate the present text or to characterize Dohrn's relationship with some people by the manner in which he met them. They show him as a human being, his high and low moods, and are also examples which clarify his practical procedures.

HOMAGE TO FRIEDRICH ALBERT LANGE

Bahrendorf bei Altenweddingen (Magdeburg)
August 30, 1866

Dear Sir,

It is not only selfishness which drives me to look for your written and, if possible, personal acquaintance. I am too profoundly indebted to you to accept calmly the unusual good I enjoyed through you, without telling you how great my admiration for you has been since I read your *History of Materialism* and how completely you succeeded in what you expressed on page 241 of your book as its purpose. In fact, your book carried me away and I must admit that never before have I met with such a complete fusion of noble humanity, sharp clear thinking, extended knowledge and harmonious beauty of expression as in this book. I had been a materialist, profoundly convinced of the alleged truth of my conviction which I considered proof against attacks. I am destroyed by your book. But instead of burdening, like a corpse on the philosophic and scientific battleground, the environment and its inhabitants with miasma and decay, instead of throwing anathema on all speculation, as most »exact« scientists still do, it seized me powerfully and led me to a complete cleaning up of the old bad growth, as it is my easily inflammable and enthusiastic nature. I let your book completely penetrate and recreate me.

But I forget, that I must at least show you my nationality and my passport so that you do not consider me a scientific hobo. I am by education a zoologist. As a high school student I already absorbed materialism through Carl Vogt, and I am attempting to acquire scientific knowledge of universal extension in order to apply for a professorship in Jena for the time being and to give lectures. My father is a wealthy private man in Stettin, whose name you may have encountered in connection with literature or music; my brother-in-law is your neighbor Wendt, director of the gymnasium in Hamm.

Your book has completely convinced me that it should be possible to combine the separate points of view of many able men of scientific and practical socio-political life on the basis of a tolerance which no longer burns or tries to eradicate men with different views. I wish this would happen soon and fast, so that the young generation which can still be formed spiritually will receive this stamp and will leave off superfluous fights about dogmas which cannot be proven; instead, it should appear as a conscious fighting force for human freedom against the dark forces. I have already tried to make your book generally known by announcing and reviewing it in a large newspaper; however, my essay did not seem to me to achieve its purpose and I will have to work it over. In the meantime, I am promoting your book in private circles and have now found in Adolf Stahr, my intimate mentor and friend, a very happy reader and admirer, such as I am myself.

Against the solidarity of all »obscurorum virorum« [dark men] the fighters for freedom and light must stick together, help and further each other wherever possible. Attribute it to these attempts and to the desire to meet you that I am turning to you so candidly and am asking you so immodestly to name all your writings to me. I have the lively desire to get to know all, particularly the ones pertaining to political economy which will reach me while I am studying Carey.

I am including a picture of myself and dare to ask you for one of you. If I may hope to be honored with an answer and to receive permission to write to you again at some time, a photograph substitutes somewhat for the lack of personal acquaintance, which of course remains my strongest wish.

In deep veneration which comes from my heart,

Yours gratefully,
Dr. Anton Dohrn

IDEOLOGICAL DISCUSSION WITH ERNST HAECKEL

Hamburg, early June, 1867

Dear Haeckel,

That you would reject the position here was my opinion from the beginning – I said this to Meyer weeks ago. Nonetheless I am bothered by the idea that my last letter could be more of a motive than I had assumed when I wrote it. I was in the strangest situation and am still uncertain about what will happen, because the decisive authorities have divided opinions. I hope you can assure me that no consideration concerning my person moved you to reject a position which seemed otherwise acceptable. You have sometimes told me the truth, when it hurt me very much – today, I hope you will not hold it back out of any consideration.

That Jena seems uncertain, I do not believe. It is part of human nature to like to talk of dangers which basically are not even believed by those who talk. And that anyone would attack you I believe even less – it would be too anachronistic. Also, you will find your loneliness more bearable – I rather fear for your peace of mind through the necessary change of mind which is before you in your views. You may remember that I told you before your departure that morphology would be an obstacle for you which would hinder you at every step – I am more than ever convinced of that. You write »how I will return to dualism is a complete riddle to me«. Indeed you should not do that. But monism can be of two kinds: consistent materialism and consistent idealism. *You* and *your* monism are *completely* materialistic, even though you quote the words of Schleicher, whose logical value is very doubtful. They are on page 105 of your first volume. Allow the following consideration to enter your mind: the contrast of mind and nature, content and form, essence and appearance is so terribly certain in every way of thinking (except in transcendental idealism) that only a desire of powerful minds is needed to wish to avoid it. To go beyond it is not only the desire of modern times, but the philosophic desire of *all* times. The materialism of antiquity arose from this desire – so did the radical idealism of the English Bishop Berkeley. Kant is the person who was most deeply convinced of the necessity of making every possible effort. It was his action (which has been compared to that of Copernicus) of turning around the relationship between subject and object and of proving that things do not determine our perceptions but that our perceptions determine the things. If I am not mistaken, this essential point has not come close to you – otherwise you could not have said: the building of critical philosophy would have been different if Kant had known the theory of evolution. You are so little correct in this somewhat arrogant sentence that even your presupposition is directly refuted, *because Kant knew the theory of evolution* and was the *first* to express it. You speak of the *Critique of Teleological Judgment* – did you read it entirely and thoroughly? Or did you take only fragments from a history of Philosophy? If the latter is the case – which I think more than likely – it is understandable that you missed the two remarkable pages in which Kant expresses the theory of evolution and adds the following words which ought to glimmer like diamonds (page 312 of the Rosenkrantz-Schubert edition): ». . . allows a somewhat weak beam to drop into the mind, that here something like a *principle of mechanism (!) of nature might be employed, without which there can be no science.*« If you had read that page, how happy you would have been! Aside from this, there are many related passages in the essays on human races in Kant. In my discussion of your *Morphology* (which died as an embryo), which Cotta's Quarterly wanted to publish, I mentioned all this. *At any rate* I shall use the first leisure time which the next years will allow me to open the door to my thoughts on Darwin as well: I am just waiting for Snell's forthcoming publication.

You blame people for not daring to think consistently: don't you suspect that one thinks and speaks apodictically only as long as one has not been radically and consciously mistaken? When, in a few months or a year you have left behind you your present views, will you then call it courage, rather than also carelessness, that you expressed such radical errors? You judge very hard and fast – I have experienced it – and you don't consider that there may be other sequences of thought which in themselves are just as consistent and justified as yours, which one can acknowledge but does not have to share; which, however, to the individual who holds them are just as reasonable as yours are to you. Is it only courage if one expresses completely opposite opinions harshly, and only cowardice if one proceeds with forbearance? See, I liked you so much and still like you as much, although I was always aware of the ever deeper and more far-reaching differences of our natures. Should I esteem you less because you do not share certain opinions? Be gentle and accept the good sides of all people gratefully and tolerate their weak sides – only then will you be like the ideal of all humanity – Goethe. I will be hanged if you do not study Lange's book to your infinite advantage. I am already a friend of Lange, and we have a wonderful correspondence. He is a marvellous person. He converted me completely – I had been like you, but I became free only through him and understood only now how powerful the fields are which cover materialism, I found the right basis in politics and social problems – truly, *I am free only now*.

I read Wundt's criticism: in many points I totally agree with him; on teleology I could have written some things with the same words he uses. I am sorry about the hint at the passage in your preface concerning inheritance and adaptation. He misunderstood you. You must not be surprised that there is such sharp judgment – you have judged much harder. And Wundt is a well-armored philosopher; I own all of his works and like to read them. You quote his »lectures«, etc., as examples of monistic investigations. Yes, but they are not monistic in your sense. Look at the motto on his contributions to the theory of sense impressions. It says: »Nihil est in intellectu, quod non ante fuerit in sensu – nisi intellectus ipse« [Nothing is in the mind that was not first in the senses – except the mind itself]. That's where the crux is. Wundt is a follower of Kant with a tendency to the philosophy of identity.

Whether I can meet you at Pentecost in Berlin I do not yet know: it depends on the turn of the situation here. Would you be so kind as to write a confirmation of my ability for the position here? It is not a matter of what I have done – that is little – but what I *can* do. I am better prepared for a position like this than for a professorship – and better than a professor – à la Claus. Goodbye and try to be my good friend again. I have not lost my hope for it. Both Meyer and Möbius send their greetings. With all my heart,

Your Anton D.—

DU BOIS-REYMOND IS TOLD THE PLAN FOR THE STATION

August 2, 1871

Honored Sir,

Although I am unknown to you, I have gathered the courage to send you the following request.

For a year and a half I have been trying with great energy to erect a large biological laboratory. I plan to build it in Naples. As often as the natural scientists ask the various governments to offer their help, they receive a negative answer because the building and maintenance costs are felt to be too high. Then I had the idea of creating the means for an institute by combining an aquarium with the laboratory. I went to Naples and, after four months of negotiations with the city authorities, I was assigned, without cost, the only adequate lot in the Villa Reale directly on the sea, under certain confining conditions. It represents an area of 7000 square feet.

On this I want to build a large building which is expected to contain, downstairs, a large aquarium open to the public, and on the upper floor 20 to 24 bedrooms and work rooms for scientists. With the income of the aquarium the expenses for the entire institute will be covered, and there will also be a considerable surplus which is to be used, not for dividends, but exclusively for scientific purposes. If it is smaller than calculated, it will still be enough to furnish the Neapolitan Institute completely with morphologic-embryological and physiological laboratories, so that these sciences can be handled there in a comparative fashion; if the surplus is considerable, the foundation of smaller stations at different coasts is planned.

The capital for the Neapolitan building (25 000 Thaler) will in large part be put at my disposal by my father, who lives in Stettin under wealthy conditions. Still remaining, however, is the procurement of 15 000 Thaler. Of this I have definite hopes for 5000 to 7000 Thaler, so that only 8000 are not yet covered.

It is my intention – and almost all financiers suggested it – to treat the whole matter as a scientific company from the beginning. Everywhere I have asked for money people have asked that the great authorities of science express their sympathy for the matter. If that could happen, the still missing finances would soon be found; if it does not, the means will simply not come.

Therefore, I have already turned to a number of my personal friends and received professional opinions and very positive writings from Darwin, Huxley, Leuckart, Haeckel, Czermak and some others.

But I would very much appreciate if I might receive from you and Professor Helmholtz something similar; and I will even dare to ask you if it would not be too inconvenient for you to allow me to let this letter be followed by an oral discussion of the circumstances of my enterprise. If it should be possible for you to give me fifteen minutes in the course of tomorrow (Thursday), I would be very grateful and you would perhaps get the impression that this favor was given to a young man who is determined to serve, with all he has, his foremost task, science. To put this plea before you is the main reason for my present stay in Berlin, which I am leaving again Friday to go to England. I hope I have not asked in vain.

I sent an almost identical letter to Privy Councillor Helmholtz.

Asking for a friendly response I am, in honest admiration,

Yours,
Dr. Anton Dohrn

AN INTIMATE LETTER TO GRANT

Berlin, Hôtel d'Angleterre
October 15, 1876

My dear Grant,

[. . .] My problems advance in more and more complicated ways. The most important event was an invitation for tea with the Crown Prince, immediately after my return from Southern Germany. It was a very small group – the Councillor from the Foreign Office, von Philippsborn, furthermore the art historian Lübke, and I. It lasted two and a half hours, first at the tea table and then in a conversation almost tête à tête; with both the Prince and the Princess I conversed for half an hour each. Both are very charming. She talks a bit of nonsense, but became enthusiastic when I talked of Adolf Hildebrand whom she declared to be the foremost living artist, something evidently imposed upon her by Hillebrand. He spoke very openly about the Bourbons in Naples, and asked about the Italian political situation. With his warmest participation he asked about the Station and begged me three times to ask him under all circumstances for support during the transfer to the Reich. I told him about the resistance of Bavaria and everything that was in my way. He obviously thought about it and decidedly took my side. »Ring at my house, as soon as you need me.« With these words he pressed my hand when I left. The Crown Princess – oh innocence – asked for Lübke's and my autograph in her album. L. wrote into it the following great sentence: »The true in the beautiful, the beautiful in the true.« I felt like writing after it, »the yellow in the green, the green in the yellow.« I did not do it, but rather wrote my scientists's motto from the *Trilogy:*

Betrachtet, forscht, die Einzelheiten sammelt,
Naturgeheimnis werde nachgestammelt.

Immediately after that evening I worked on the definitive request to the Office of the Chancellor, a work which took three days of corrections, etc.; I presented a copy to the Crown Prince and had the satisfaction of knowing that the secretary, Mr. von Normann, said two days ago that the Crown Prince had immediately written to the Chancellor's Office and asked that »my request be treated as his own«. So far, so good.

But now listen. During the Hamburg Meeting of Naturalists, Kossmann, supported by a clique – Semper, Pagenstecher, etc. – made the proposal that the German Reich should start zoological stations in Helgoland and Kiel, and should promise 100 000 Mk. at once and 20 000 Mk. annually. An editorial committee (Haeckel is on it) for the petition has been chosen and hopes to proceed fast. »Nobody wanted to go against the beautiful institution in Naples, etc., etc.; but for distance, health conditions, etc. – Mayer brought me the news. Fortunately it hits me here, not in Naples, so that I can take immediately the great firehose, and I hope to have extinguished the fire before it spreads. I immediately denounced the matter at the highest place in such a way that it does not compromise me and I evoked a definite expansion of the forces on my side. I also urged a quick support for my request so that it would be passed during the November session of the Reichstag. In a short while I will try to start a counter-movement in scientific circles – and at least delay the preparatory steps of the committee. Gaining time is everything. Hensen calls the matter »harmful and silly«. Others, too, are against it.

You see, the gang tries to explode me. But there are countermines. In their proclamation, they complimented »the munificence of the German Reich and the individual States, which made it possible to work in Naples«, but my name is not mentioned.

I hope the whole group will feel that I am enough of a man to ward off this blow. If only I were healthy, I could not care less, but I am no longer sure of my strength. However, I am certain that in case of need I will have the most direct intervention from above – but such a trump has to be avoided as long as possible.

In the middle of this diplomatic game of chess my moods of course vary as always. Marie is far from me, so are you – and all other human beings have no souls, as far as I am concerned ... [here follows the passage on Fanny Lewald rendered on p. 43].

Marie writes the following on Tom Jones: »I have finished T. J. and enjoyed it to the last, indeed will often enjoy it in thinking over it and remembering the comical situations. One is quite carried away with the keenness of life in it and as to me personally I do not think it has harmed my soul. It has perhaps awakened more insight into the real motives of many human deeds, motives which still now were unrevealed to my pure mind. In this T. J. has done the work of some years of life.«

And as I write this, one of the great pains begins to bore in me, which always overwhelms me when I am just ready to express the entire fulness of my love of her which I, from one day to the next, still feel in its growing dimensions. Dear Grant, I, too, know the bloody spots, which make the most beautiful edition of a great work often illegible – and I may try as hard as possible to read over them or to erase them – if I turn the page drops fall on it immediately.

I have smoked a cigarette and changed into blue haze what just threatened to block my way in order to chat with you again. That Dew (Smith) has not written is very much like him – but I am convinced that he does not understand how it hurts.

Unfortunately I cannot send you any money at the moment, I had to send Marie everything I could get; her father is short of money because of Serbia, etc., and she needs money for the birth of the baby. But borrow wherever you can – I will work us free for a long time. Oh God, yes – I am so anxious to live with her and with you and with our children – who are also and will remain yours – that is certainly true. The boy is as charming as his mother, angry and tender, always full of an enterprising spirit and so happy.

I decided to give a slavic name to the second child also. These are new declarations of love I make which will last when I am no more. I am now reading Polish history and I read it with my heart. When, the other evening, I listened to music by Chopin, I felt as if Polish blood was working in me, the entire wild lack of control, the insane knighthood, the grand criminal power – I saw the old Pole with the sharp nose kneel before Countess Potocka in Krzeszowice and revere her like his queen – and then the spurs clatter to the Mazurka – forgive me, Grant – I am awakening all your pains – but they are a lovely people, and they had to perish . . . Here, where I allow only the image of the nation with its glitter, and where the word Poland holds me on to Marie with hundredfold chains, I can for quite some time let the feelings remain which teach me what those feel who love Poland, who are Poles. That is why my children and hers are going to have Polish names.

Goodbye, dear old friend. I am working on re-conquering you and giving you again that peace of mind which got lost in your thoughts and feelings, that you were a piece of the material difficulties of life for me. Yet you knew and must know that the contrary is true! The three chains which Marie, Bux and I gently fastened to your arms cannot so easily be stripped off.

Once you have convinced yourself of it during the winter, your room in Palazzo Torlonia will again be well furnished and Bux will reclaim his little corner in it.

Goodbye, wholehearted greetings to your hosts.

The next letter will report the birth of the child.

Yours, A. D.

THE PLANS AND THE QUESTION OF THE CALL TO BERLIN

May 17, 1883

Honored Sir! (du Bois-Reymond)

Your long letter has moved me in many respects. First of all, I want to thank you with all my heart for cutting out of your short amount of time such a large part in order to write me. I already know the chapter of unanswered letters and ponder about how to allow myself a secretary who would be for me more an advantage than a weight. It will probably remain a project, such a »rara avis« is hard to find, particularly since it is a question of a correspondence in four languages. My own wife is a born secretary, but social and family duties take care of her time and energy. Therefore, it has to go without.

In fact, I read your work immediately and believe that I understood what its main concern is. I see perfectly how important this question is for your life's work: aside from the polemics in which you are engaged there is a kind of artistic sensitivity in knowing that the end of your scientific life's drama is in harmony and growing in development with the beginning. Of course, I cannot form an independent judgment of the problems, but I like to remember what Willy Engelmann told me some time ago: whatever the result of the great electro-physiological debate, your work will remain of the same epoch-making significance and all of modern physiology rests on it.

Recently I had the opportunity here of meeting the hasty and immature Grünhagen: we were freshmen together in Königsberg and a kind of acquaintance results from this. However, the impression he now made on me was eminently insignificant; above all he seemed to me constantly out of breath intellectually. How he expects to climb a large intellectual mountain, I do not know. He wanted to make some investigations here on the torpedo, but knew so little of the anatomy of the animal that I gave him an Italian student to help him find the vagus.

The matter of the physiological laboratory of the Zoological Station has gone through a development which, although delaying it, also promises to enhance its size and intensity. The originally considered building proves to be hazardous to health. Alterations would be unavoidable, but if some construction must be done it will be more profitable to erect a new building next to the Zoological Station, provided the Italian government will give the needed

50000 frs à fonds perdu. My present task is to achieve this. I hope I will succeed. Then space would be gained, and we could achieve a real collaboration of physiologists and morphologists.

Soon I must go to Rome to get a definitive answer from the secretary. Only then can I undertake any further steps, including the request to the Academy. I hope this delay does no harm to the result.

I believe that the workings of the chemical laboratory will be expensive, but I have already required the zoologists to be more careful with their use of alcohol by making them pay for alcohol beyond a certain quantity. I would do the same with any chemical substances which we might supply. Everybody should find out where to get the needed money.

I am glad that you are thinking of me in relation to the question of filling the position of the zoological chair at the Berlin University. If your opinion were correct that, concerning the Station, all difficulties are overcome and I can leave it, the position in Berlin would indeed tempt me, not despite the museum, but really for its sake. I would be very tempted to break with all the old museum plunder and to put in its stead a new organization whose shape I see rather clearly. But I have not yet attained here what is attainable and what, after all, is more propitious to the problems of zoology than an ever so successful teaching position. We still move too exclusively within the limits of comparative anatomy and embryology: comparative physiology and the study of the way of life still await a forceful initiative, which can best be achieved here.

To be sure, the eventual occupancy of the Berlin professoriate interests me very much. Haeckel would certainly be the right man, if he weren't such an overgrown child . . . There is no doubt that Weismann is intellectually the most important person since Siebold has become decrepit, but he is very nervous and distrustful and, I believe, a very uncomfortable colleague. Eilhard Schulze is a nice man, a good zoologist, but without great distinction. If it could be done without hurting German national feelings, I would put Mečnikov, who has no position right now, on the list; and Kleinenberg in Messina is also a very excellent scientist. The main difficulty, however, seems to me to be in the combination of the museum and the professorship. It could, as far as I can see, be solved only *if the Department of Education would decide to separate museum and professorship*. Nobody can serve two masters. The goals of the museum are, if correctly understood, totally different from the professorship. The museum would have to be submitted to a man like Möbius (Kiel) or Ludwig (Giessen) or von Koch (Darmstadt). For the moment one would have to limit oneself to employing a custodian or conservator who would only see to it that the animals would not perish materially, i.e., non-professional assistants – the head would have to use the wages of the presently employed gentlemen or newly allowed sums to select specialists who would be entrusted with the elaboration of individual groups. The purchase of new collections, etc., would have to be stopped: work at the museum must grow in intensity rather than breadth. The presently popular catalogues are silly: small monographs in the style of the Flora and Fauna of the Gulf of Naples are needed. The head of the museum would have to engage young men, let's say half a dozen at the same time; each of them would have to work on a separate group and each would have to discuss with him principles of classification and would have to follow his direction. Then systematics would gain a connection with genuine morphology which it now lacks and would stop shaping the species and genus manufacture to become a mania and a plague.

If it were possible to move the government toward such a separation of museum and professorship, I would like to offer myself to take over the building, organization and directorship of the museum. I would spend the winter in Naples and the summer in Berlin, until everything had been put in place (approximately five years) and then the Museum could be permanently handed over to someone else. I would enjoy this and it would fit the ideal of my activity, which I conceived thirteen years ago at a moment of a certain boldness and of which I have already realized a fairly important portion – diis et amicis faventibus [favored by gods and men]. If Goeppert were still alive, I would perhaps have tried to propose such an arrangement, but now I lack the personal contact. Perhaps you trust me enough to suggest to Gossler that he should ask me, first of all, for my opinion on the situation? I am arrogant enough to consider myself capable of pouring new blood into the veins of the museum and with it into systematics as well. Concerning the financial side, I have learned enough here; I am also practiced in dealing with human beings.

If this separation of museum and professorship could be decided on, the appointment of a professor would be incomparably easier. Weismann, or Mečnikov, Kleinenberg – these three as outstanding zoologists and also as high class human beings can be named in the first place. Soon enough younger men will crowd around them and »Privat Dozenten« would quickly complete the zoological school of the University of Berlin. As an Honorarium Professor I would like to lecture during the summer on morphology of articulates or morphogeny. This would not cost me much trouble. [. . .]

If at the end of this long epistle I may express a wish, it would be the following: to make public, in some way, the fact that I was considered for the nomination of the Berlin professorship. Whether the only way in which I could combine my duties here with an eventual nomination could be considered, I do not know. But for my further plans it would be of great help, if in a decent way the question of a nomination could become public. I had already planned to ask if you could write to Blaserna or Todaro and point out that in Berlin all circles counted on Italy to put up some money in order to further the goals of the Zoological Station and that you hoped it would correspond to my wishes to find means for the physiological laboratory. Virchow intends to talk about it to Baccelli – whether he has already done so, I do not know. If you and perhaps Helmholtz would intervene in this sense (perhaps directly with Sella or Baccelli) I believe that the matter would not encounter difficulties. I would present the potential nomination to Berlin in the right light and thus produce the needed pressure.

That last year's increase of 10 000 Marks of the subsidy by the Reich did not materialize is Bismarck's fault. But I have such a good relationship with his secretary, von Boetticher, who lies here sick, that he promised to use his influence to remedy the matter. There is also a prospect for further help among private circles, so that I hope to complete my program entirely.

And now, last but not least: If all warnings are obeyed, if rest is observed and dietary mistakes avoided, if travel by sea rather than land (Genoa-Napoli) is chosen, if the little steamer is taken directly from the boat which brings you all the way from Genoa to Naples, and lands you in Casamicciola where you will slowly learn to tolerate the heat, the trip may be taken without danger. I suggest the following distribution of time: two weeks in Casamicciola, two days in Capri, two to three days in Amalfi (Ravello, boat trip along the coast), two days in La Cava, from there a round trip to Pompeii, a trip to Sorrento, one day there; (Baja perhaps, from Ischia by steamboat), and finally Museo Borbonico, Aquarium, Camaldoli – then to Rome, where you have to live with caution in the Hotel d'Angleterre, which is healthy – and your parental heart will be able to return in peace to Berlin. But caution when you enter Italy: no fruit, little water! No strain, neither physical nor psychological, trivial food, beefsteak and cutlets, moderation about wine, much soup – thus you will keep the young ones healthy and active. I am not a natural philistine, but in this matter philistine behavior is best. I can and will take care of cheap boarding houses.

Yours, today and forever,
Anton Dohrn.

My wife claims the forgotten greetings and I hurry to add them.

FEELINGS OF VICTORY AND CONFLICTS AT HOME (TO HIS SISTER)

March 12, 1884 evening
[Naples]

Dear Anna,

Since I plan to leave in about a week and I know you are in Stettin, I will tell you about the situation and my opinion of it. The fact that I am not yet in the Hotel Kaiserhof where your letter to Marie from Magdeburg assumes me to be is connected with my campaigns here. I am just about to gain another great victory. The Zoological Station had in its original program the addition of an area for physiology. I did not, however, reach that goal in the first ten years. Zoology proper has grown enormously thanks to the Station, and has had so much to feed on thanks to the ideas of Darwinism that it occupies the entire large building completely and more than completely; even for the traditional purposes the Station has become too small.

You know that Bismarck will not grant me new means. Whether his opposition will last is doubtful; it would be dangerous to put it to the test just now. Furthermore I am bound to maintain the Station »in its present condition« for the duration of the subsidy of the Reich. The only thing left to me is to find means for the extension from some other place. For two years I have worked to win Italy. After much effort this now seems to be succeeding in a surprising fashion. I have spread my net up to the King, and I believe that all birds of importance are now in it. Italy will pay approximately 100 000 francs; I will build a large new building immediately next to the old Station building, and Italy will also pay the lion's share of the new daily expenses. It will be decided within the next few days.

This makes an immense achievement possible. What Darwin produced through his individual books, each of which created almost a new discipline, I succeeded in doing by creating new organizational turns. The range of the one started now is very large, and the superiority which the Station will gain by it will be decisive for all future projects.

But the matter has still another, political, side. Up to now I really had only Germany on my side: the others were angry or followed reluctantly, and Italy was not the least among them. Now I am about to form a close alliance by holding together the means and power of both nations. Whoever up to now looked at the Germans askance must now willingly recognize the international character of the Institute. The Italian half or third will prevent the possibility that the Station could be taken away from me as being a German Reich Institute. Inversely, Italy can never push out Germany – thus I, Anton Dohrn, and my sons will remain indispensible connecting links. Instead of using one horse, I will travel with two. If one becomes lame I can use the other. This is a far-reaching example, but the whole enterprise is far-reaching and does not have its equal.

As soon as I have fixed things here, I will go to Berlin in order to take over the subscription for the new steamer with the help of »Fritz«. For this, the secured triumph in Italy will be most helpful. Once that has been successful, I will go to Petersburg, Moscow, Stockholm, London, Holland-Belgium, in order to secure their following for a long time and to confirm old friendships by my personal appearance, to win new ones and to disarm sneaking enemies.

In this the fortunate ending of my long fight with Gegenbaur will be of great assistance. In fact, Gegenbaur is annihilated. You know with what perseverance I have prepared for this moment and you can imagine what consternation has seized all of those who considered him a god and me a visionary boy. I could show you letters from the most competent judges which would elate your sisterly heart. Believe my word: the great man is Kaput. One single well-aimed shot was enough; and that he will not rise again – future arrows of which I have plenty in my quiver will make it certain. For years to come my dominance over biology is secure, theoretically as well as practically – I have become, in a certain, not immodestly meant sense, Darwin's executor.

So much for characterizing the present phase of my scientific career. The contrast could not be greater if I now emphasize my relationship to Lindenstreet. CAD has instructed me through Heinrich, as if I were a dumb boy, that I am a miserable maker of projects who completely misunderstood the »modus vivendi« [way of life] when I took upon my shoulders Marie's intention to ask him for 10000 frcs., because I was unwilling to hide behind her apron. The »modus vivendi« must therefore now be taken seriously by me: its contents can only have been that as long as mother lived, I was permitted in the parents' house, i.e., as the son of my mother and father I was not to be pushed out. Mother is dead – the »modus vivendi« is settled. I draw this conclusion.

I am determined to let such an explanation reach Stettin when I am in Berlin and, with all consideration in my expression, clearly declare that I will not set foot in father's house in Stettin or Hökendorf except with an expressed invitation. I have to go to Stettin because of the Vulcan, where the machine for the next steamer is being built.

This is the situation which you may consider as being firmly decided on my side. If a reconciliation is possible, I do not reject it – but you are the only one to bring it about and I can undertake it only without any further humiliation. This, too, you will have to hold on to.

Goodbye and trust your brother – he can continue to exist without father's financial crutches and cannot be imposed upon to surrender unconditionally.

Yours,
Anton

Greetings to good old Jette and Pitsch, also to Miss Rahtz.

ON THE DEATH OF ERNST ABBE

Naples, February 2, 1905

Esteemed Friend, (Elise Abbe)

If I am sending you a few lines only today, you know without my assurance how this is meant. I need not tell you what moved my mind when I heard of the final deliverance of the poor sufferer; in me lives the memory of those times in Jena when we both were growing and experiencing so many things in common with you and your father. That memory is as bright as ever. Only you, who lived with us through those years, know what they meant to us. When I now read the obituaries which finally acknowledge before the entire world the rank of a great personality, I feel as if I can say: now, after more than thirty years, you others finally understood the man, in whose mind I knew decades ago all the germs which later developed to greatness, but which did not add to him as a man anything that was not already in him at that time, but which was not regarded as possible by anyone except our small circle.

It's true that such an immense realm of activity as he created for himself also brought other dimensions to his personality, but the nobility of his person was already the same, when he was still living in the little room in the Saalgasse and we ate our warmed-up meal together in the little arbor or took a walk with you and Papa Snell toward the meadows of Löbstedt or Wöllnitz. I thought in a more lively way than ever of those years and told those who share my house and my friends at the Zoological Station many details.

Thus I was close to you in my thoughts and knew how you would have helped my memories – as at the time when the two of you were here with me for the last time.

»Oh, give me back the times when I was still a growing man.« We all were still growing at that time. He and I became someone, and both of us placed great goals for our strength and reached them. How much greater he always was, was shown by the goals we reached. Grateful for the loyal friendship of the great friend, I may tell you how I look up to him and revere him as the greatest person I knew. Now the earth covers his mortal frame and now he is at rest – how he longed years ago to finish his great theoretical works! This did not happen, and it is a serious loss for the world and for him. I know myself that only *one* great goal still stands before me, namely to finish up *my* world of thought; and I can therefore judge how he, too, was dominated by this wish. Then came the dusk – and the night; and now everything is over . . .

In old loyalty,
Yours,
Anton Dohrn.

NOTES AND REFERENCES

1 The genealogy of the Dohrn family has been researched by Dr. Bethe (Stettin); he published his results in *Familiengeschichtliche Blätter* [Pages of Family History], 1938, issue 9–10.

2 O. Altenburg, *Pommersche Provincial-Zuckersiederei* [Pomeranian Provincial Sugar Refinery] *1817–1917*. Memorial to its hundredth anniversary. Stettin, 1917.

3 The city library of Stettin (Director Erwin Ackerknecht) started a major collection of the correspondence of C. A. Dohrn in original and in copies. According to a remark of Marie Dohrn, Heinrich the younger destroyed much of the family correspodence after his father's death. Yet a notebook of his trip to Brazil was kept. Part of his youthful letters have been used in the longer biographical study which Heinrich Dohrn published in 1892: Carl August Dohrn. *Stettiner Entomologische Zeitung*, 53 (1892) 281–322. Also, C. A. Dohrn's niece Marie Grotjohann, who lived in his house during his old age, wrote down some remembrances of C. A. Dohrn. However, the most important source are the 1840 to 1892 volumes of the *Stettiner Entomologische Zeitung* with their numerous autobiographical notes. C. A. Dohrn's entomological work is described in W. Mayer, Geschichte des Stettiner Entomologischen Vereins [History of the Stettin Entomological Society]. *Stettiner Entomologische Zeitung*, 100 (1939) 2–71.

4 See: Heinrich Dorn, *Aus meinem Leben. Musikalische Skizzen* [From my life, musical sketches], Berlin, 1870.

5 Graf Schack, *Perspectiven*, vol. I, 1894.

6 Some of the poems which C. A. Dohrn had sent to the King are kept in the Royal Archives of Brandenburg-Prussia.

7 Gustav Kraatz, *Ein Beitrag zur Geschichte der systematischen Entomologie* [Contribution to the History of Taxonomic Entomology], Berlin, W. Horn, 1906. This also treats the conflicts between Kraatz and Dohrn.

8 *Bref till Adolf Fedrik Lindblad*. Stockholm, 1913. The friendship with the Swede who owed his popularity at that time in Germany largely to Dohrn, had probably been started by Felix Mendelssohn.

9 Protocols of the Prussian House of Representatives, 1859–62.

10 Aside from the German entomological journals the following professional journals published an obituary after C. A. Dohrn's death: *The Entomologist's Monthly Magazine*, (London) 28 (1892); *Transactions of the Royal Entomological Society of London*, (1893); *The Canadian Entomologist*, 24 (1892) 181–182 (August R. Grote); *Entomological News*, Philadelphia, 3 (1892).

11 The letters written to Haeckel by Anton and Carl August Dohrn are kept in the Ernst Haeckel Museum in Jena.

12 Letters and documents on Heinrich Dohrn are in the City Library of Stettin. See also: W. Mayer, (1939; see note 3).

13 A typical selection of Stahr's correspondence was prepared with an introduction by Ludwig Geiger (Berlin, 1903). For Fanny Lewald as a personality and representative of her time, her *Lebensgeschichte* [*Autobiography*] is essential (Berlin, 1861–62).

14 On Carl Gegenbaur: Max Fürbringer, Carl Gegenbaur. *Naturwissenschaftliche Wochenschrift*, 7 (1903) 103–109; E. Goeppert, [Carl Gegenbaur], *Das Biographische Jahrbuch*, 8 (1903) 324–339.

15 Otto Adolf Ellisen, Lange's biographer, could already use the Dohrn-Lange correspondence in his book *Friedrich Albert Lange. Eine Lebensbeschreibung* [A Biography], (Leipzig, 1894). With minor shortcuts the entire correspondence was reprinted in 1933 and evaluated from the viewpoint of intellectual history: Rudolf Carnap & Hans Reichenbach, editors, Dokumente über Naturwissenschaft und Philosophie. Briefwechsel zwischen Friedrich Albert Lange und Anton Dohrn [Documents on Natural Science and Philosophy. Correspondence between F. A. Lange and A. Dohrn]. *Erkenntnis*, 3 (1932–33) 262–300.

16 In *Life and Letters of Thomas Henry Huxley*, edited by L. Huxley, (London, 1900) some of the correspondence of the two men is used.

17 Adolf Hildebrand told about his early relationship to Charles Grant in his unfinished *Lebenserinnerungen* [Memories of my Youth], *Süddeutsche Monatshefte*, (1931).

18 Among the numerous studies on Ernst Abbe, the biography by Felix Auerbach remains the standard work: *Ernst Abbe. Sein Leben, sein Wirken, seine Persönlichkeit, nach den Quellen und aus eigener Erforschung geschildert* [Ernst Abbe. His Life, his Work, his Personality, Compiled from Sources and from Personal Research], Leipzig, 1918. (Große

Männer, 5). The physics achievement has been discussed by M. von Rohr: *Ernst Abbe*, Jena, 1940. The socio-political development of Abbe has been elucidated by Friedrich Schomerus: *Werden und Wesen der Carl-Zeiss-Stiftung* [Nature and Development of the Carl Zeiss Foundation], Jena, 1940.

19 According to the records of the Friedrich Schiller University in Jena, Dohrn had sent five theses from Montreux on April 25, 1868, for the »debate«:

1. The application of Darwin's theory on man is contrary to the consequences which Malthus drew from his laws of increase of population and nourishment.
2. Darwin's provisional hypothesis of *Pangenesis* is not sufficient to explain the phenomena of heredity.
3. Paleontology is of little value for the determination of genealogy in the animal kingdom.
4. On the contrary, the comparative developmental history is of invaluable service for the detection of the genealogy of organisms, since the developmental history of the individual repeats to a large extent that of the phylum.
5. The number of German universities must not be reduced.

20 On the relationship Darwin Dohrn see also: *The Life and Letters of Charles Darwin*, edited by Francis Darwin, 3 vols., London, 1887 [reprint, 2 vols., New York, Basic Books, 1959].

21 Anton Dohrn, *Der gegenwärtige Stand der Zoologie und die Gründung zoologischer Stationen* [The Present State of Zoology and the Foundation of Zoological Stations], *Preussische Jahrbücher*, 30 (1872) 137–161; reprinted with complementary footnotes in: *Die Naturwissenschaften*, 14 (1926) 412–424 (special number for the 75th birthday of Adolf von Harnack).

22 For Grant's life and personality see J. B. Capper's introduction to the posthumous edition of his book *Stories of Naples and the Camorra*, London, Macmillan & Co., 1896, pp. v–xvii.

23 The development of the Naples Frescoes is discussed in great detail in: Julius Meier-Gräfe, *Hans von Marées. Sein Leben und Werk* [Hans von Marées. His Life and his Work], 3 vols., Munich, 1909–10.

24 During the first decade and a half, Dohrn reported regularly on the development of the Zoological Station in the *Mittheilungen aus der Zoologischen Station zu Neapel;* these activity reports were used, together

with his letters. For the later period the main source were his annual reports to the State Department in Berlin.

25 Johannes Reinke talked about his stay in Naples in his autobiography *Mein Tagewerk* [My Day's Work], Freiburg, 1925.

26 For the collective volume *Aus dem Leben und Wirken von Arnold Lang* [From the Life and Work of Arnold Lang], (Jena, 1916) Hugo Eisig wrote a lengthy essay *Arnold Lang und die Zoologische Station in Neapel, 1878–1885* which gives a somewhat anecdotal image of the workings of the Station and the human relations of friendship.

27 H. V. Petersdorff, [Robert von Keudell]. *Biographisches Jahrbuch*, 1905. – Robert von Keudell, *Fürst und Fürstin von Bismarck. Erinnerungen aus den Jahren 1846–1872* [Memories from the years 1846–1872], Berlin, 1901.

28 Aside from the reports of the Reichstag, the protocols of the committees were used.

29 Aside from material from the Archive and the reports in the *Mittheilungen*, Dohrn's essay *Aus Vergangenheit und Gegenwart der Zoologischen Station* [About the Past and Present of the Zoological Station], *Deutsche Rundschau*, 72 (1892) 275–298, has been used.

30 On Nansen's sojourn in Naples see: *Vossische Zeitung*, September 17, 1896; there are various biographies on Nansen, the most extensive of which is the early one by W. C. Broegger & Nordahl Rolfsen, *Fridtjof Nansen 1861–1893*. [English translation] London, New York, Bombay: Longmans, Green & Co., 1896. [German translation] Berlin, 1897.

31 Valerian Kalinka, *Der vierjährige polnische Reichstag 1788–1791* [The Four-Year Polish Congress], Berlin, S. Mittler & Sohn, 1896–98.

32 Arnold Sachse, *Friedrich Althoff und sein Werk* [Friedrich Althoff and his Work], 1928. The conflict with Dohrn is not mentioned in this book.

33 Anton Dohrn, *Das 25-jährige Jubiläum der Zoologischen Station zu Neapel* [The 25th Anniversary of the Zoological Station at Naples], Leipzig, Breitkopf & Härtel, 1897. The most important of the essays published in honor of that day is a richly illustrated essay by the Freiburg zoologist H. E. Ziegler: Die Zoologische Station zu Neapel. Zur Feier ihres 25-jährigen Bestehens. *Velhagen und Klasings Monatshefte*, 11 (1896–97) 416–425. Dohrn refused to give his portrait for this occasion.

34 See the minutes of the annual meetings of the German scientists and physicians in Rostock (1871), Breslau (1874) and Berlin (1886).

35 In the book by Klein & Hehemann, *Friedrich Alfred Krupp.* (1903), Krupp's zoological interests and his amicable scientific relations with Lobianco are briefly treated.

36 The life and work in the Station during Anton Dohrn's last decade are discussed by Wilhelm Giesbrecht: Altes und Neues aus der Zoologischen Station in Neapel [Past and Present News from the Naples Zoological Station], *Deutsche Rundschau*, 33 (1907) 76–92; and by Hans Spemann: Die Zoologische Station zu Neapel. *Süddeutsche Monatshefte*, 4 (1907) 136–149. A special issue of the *Publications of the Association of Collegiate Alumnae* (Feburary, 1907) is also dedicated to the Station.

37 For the relationship between Wolf Dohrn and Friedrich Naumann, see Naumann's memorial speech for Wolf Dohrn, *Berichte der Dalcroze-Schule*, 1 (1914); and Th. Heuss, *Friedrich Naumann*, Stuttgart, 1938.

38 Theodor Boveri, *Anton Dohrn. Gedächtnisrede gehalten auf dem Internationalen Zoologen-Kongress in Graz am 18. August 1910.* [*Anton Dohrn. Memorial Speech delivered at the International Congress of Zoologists at Graz on August 18, 1910*[, Leipzig, S. Hirzel, 1910. 43 pp. – Of the numerous obituaries I should name: Hans Driesch, Zur Erinnerung an Anton Dohrn. *Süddeutsche Monatshefte*, (1909) 513–518. – von Haustein, *Naturwissenschaftliche Rundschau*, (1909). – Wilhelm Waldeyer, Anton Dohrn zum Gedächtnis. *Anatomischer Anzeiger*, 35 (1910) 596–603. – Ray Lankester, Anton Dohrn. *Nature*, 81 (1909) 429–431. – W. Kükenthal, *Jahresberichte der Schlesischen Gesellschaft für Vaterländische Kultur*, (1909). – F. W. Winter, *Berichte der Senckenbergischen Naturforschenden Gesellschaft*, 41 (1909). – R. Hesse, *Berliner Tageblatt*, (September 29, 1909).

Jakob von Uexküll published in his *Niegeschaute Welten* [The Worlds Never Seen], (Berlin, 1937) an essay on Dohrn. See also E. Korschelt, *Das Haus an der Minne* [The House on the Minne], Marburg, 1939; Curt Herbst, Anton Dohrn, in: *Pommersche Lebensbilder* [Lives of Pomeranians], Bd. 1 (1912) 293–303.

39 A. von Tschermak-Seysenegg, Die Zoologische Station in Neapel. *Meereskunde*, 8 (1914) 1–40. The author discusses, according to the state of science at that time, the essential scientific achievements which started in Naples.

Margret Boveri, Die Zoologische Station zu Neapel. In: L. Brauer, A. Mendelssohn-Bartholdy & A. Meyer, *Forschungsinstitute. Ihre Ge-*

schichte, Organisation und Ziele. [*Research Institutions. Their History, Organisation and Scopes.*], vol. 2, Hamburg, 1930, pp. 578–598.

Margret Boveri, Die Zoologische Station in Neapel. *Atlantis*, 5 (1933) 564–576.

H. Braus, Die Zoologische Station in Neapel. *Die Naturwissenschaften*, 12 (1924) 544–548: about the return to the directorship of the Station of Reinhard Dohrn.

In *Abenteuerliches Herz* [*Adventurous Heart*] (Hamburg, 1838) Ernst Jünger dedicates a short part *Frutti di Mare* [Fruits of the Sea] to memories of his biological work in the Naples Station.

40 A special issue of *Die Naturwissenschaften* (vol. 28, no. 51, December 20, 1940) was dedicated to Anton Dohrn with contributions by H. Spemann, F. Baltzer, U. Pierantoni, G. Cotronei, F. Pentimalli, F. von Wettstein, H. Rein, E. von Holst, M. Hartmann, F. Süffert; it also includes a reprint of Theodor Boveri's memorial speech (see note 38).

In the *Erlebnis-Bücherei* [Library of Adventures] a »factual report« was published by W. Kreuz, *Auf Krakenfang vor Neapel* [*Catching Octopus off Naples*] (Berlin, n.d.) that describes Dohrn's beginnings in Naples in a phantastically enhanced novellistic form.

INDEX OF NAMES

LIST OF FIGURES

MIX
Papier aus verantwortungsvollen Quellen
Paper from responsible sources
FSC® C105338

If you have any concerns about our products,
you can contact us on
ProductSafety@springernature.com

In case Publisher is established outside the EU,
the EU authorized representative is:
Springer Nature Customer Service Center GmbH
Europaplatz 3, 69115 Heidelberg, Germany

Printed by Libri Plureos GmbH
in Hamburg, Germany